Bibliografische Information der Deutschen Nationalbibliothek

Die Deutsche Nationalbibliothek verzeichnet diese Publikation in der Deutschen Nationalbibliografie; detaillierte bibliografische Daten sind im Internet über `http://dnb.d-nb.de` abrufbar.

Covergestaltung: Ines Melzer

Logos Verlag Berlin GmbH
Comeniushof, Gubener Str. 47,
10243 Berlin
Tel.: +49 (0)30 42 85 10 90
Fax: +49 (0)30 42 85 10 92
INTERNET: `http://www.logos-verlag.de`

Fast and approximate computation of Laplace and Fourier transforms

Dissertation
zur Erlangung des Grades
Doktor der Naturwissenschaften (Dr. rer. nat.)
des Fachbereiches Mathematik/Informatik
der Universität Osnabrück

vorgelegt von
Ines Melzer

Osnabrück, September 2015

Erstgutachter: Prof. Dr. Stefan Kunis

Zweitgutachter: Prof. Dr. Laurent Demanet

Tag der mündlichen Prüfung: 05.02.2016

2010 AMS Mathematics Subject Classification: 65T40, 65T50, 42A15, 30E10, 65D05, 65F30.

Contents

List of publications

The following list of publications summarizes the parts of this thesis which have already been published. The respective numbers are the same as in the full bibliography at the end of this thesis.

[32] L. Kämmerer, S. Kunis, I. Melzer, D. Potts, and T. Volkmer. Computational methods for the Fourier analysis of sparse high-dimensional functions. In, *Extraction of quantifiable information from complex systems*. Vol. 102, in Lect. Notes Comput. Sci. Eng., pp. 347–363. Springer, Cham, 2014.

[35] S. Kunis and I. Melzer. A stable and accurate butterfly sparse Fourier transform. *SIAM J. Numer. Anal.*, 50(3):1777–1800, 2012.

[36] S. Kunis and I. Melzer. Fast evaluation of real and complex exponential sums. *Preprint*, 2014.

[37] S. Kunis and I. Melzer. XFFT, Matlab toolbox for computing extended fast Fourier transforms: butterfly sparse Fourier transform, Laplace transform, fast Fourier transform for complex evaluation nodes. `http://sines.de`. 2015.

Chapter 1

Introduction

Recently, the handling of digital information becomes more and more important in workaday life as well as in current research. The main idea is to reduce the huge amount of data by using appropriate representations which need much less data but provide nevertheless a good approximation of the original information. Highly efficient algorithms to process huge data sets are therefore of fundamental interest. One basic problem in numerical analysis and scientific computing is the approximation of functions. A very famous and well known example is the fast Fourier transform (FFT) introduced firstly by C. F. Gauß in [21] and most popularly published by J. Cooley and J. W. Tukey in [9]. The FFT belongs to the algorithms with large impact on science and engineering. It allows very efficient computations of trigonometric interpolation polynomials to functions.

In this thesis, we consider fast and approximate algorithms to evaluate a function f given by coefficients $\hat{f}_k \in \mathbb{C}$ and frequencies or nodes $\boldsymbol{\xi}_k \in \Omega$, $k = 1, \ldots, M_2$, $M_2 \in \mathbb{N}$, of the form

$$f(\boldsymbol{x}) = \sum_{k=1}^{M_2} \hat{f}_k \kappa(\boldsymbol{x}, \boldsymbol{\xi}_k), \tag{1.1}$$

where $\kappa \colon X \times \Omega \to \mathbb{C}$ denotes a so called kernel function, $X \subset \mathbb{R}^d$ the spatial domain, and $\Omega \subset \mathbb{R}^d$ the frequency domain. A similar task is the continuous counterpart, i.e. the computation of the integral transform

$$f(\boldsymbol{x}) = \int_\Omega \hat{f}(\boldsymbol{\xi}) \kappa(\boldsymbol{x}, \boldsymbol{\xi}) \mathrm{d}\boldsymbol{\xi},$$

where Ω is an open set and $\hat{f} : \Omega \to \mathbb{C}$.

Techniques for the computation of function values of f given by the integral transform with a smooth kernel are fast multipole methods [26, 50, 59] as well as the concept of hierarchical matrices [28, 3, 25, 4, 29]. In all cases, the concept in such schemes is to trade exactness for efficiency; instead of precise computations up to machine precision, the proposed methods guarantee a given target accuracy. Neglecting logarithmic factors in the problem size and the target accuracy, the computational complexity of all these algorithms scales linearly in the problem size. The idea of all above cited works relies on a simple geometric condition for subsets of X and Ω where the kernel κ can be approximated by a low rank approximation restricted to these subsets. We call these geometric conditions admissible conditions and call the subsets admissible. Subsequently, we deal with different kinds of exponential kernels. There are other methods for the computation of fast function transforms which rely on asymptotic expansions of functions as presented for example in [30, 51].

In the first part of this thesis, we study the univariate real valued exponential kernel $\kappa(y, \xi) = e^{-y\xi}$ which leads to a discrete Laplace transform. This transform has been studied in [46, 49]. In the latter, the author uses a particular function approach by Laguerre polynomials, whereas in [46] interpolations are used. We introduce the discrete Laplace transform [46] in a modern form including a generalization to more general kernel functions and an improvement in the error estimate. These more general kernels lead to specific function transforms. We present a complexity analysis of the generalized version and illustrate the theoretical results by numerical experiments. All these findings can be found in our preprint [36].

In the second part of this thesis, we consider the Fourier kernel $\kappa(\boldsymbol{x}, \boldsymbol{\xi}) = e^{2\pi i \boldsymbol{x} \boldsymbol{\xi}}$ with real variables $\boldsymbol{\xi}, \boldsymbol{x} \in \mathbb{R}^d$ for nonequispaced nodes which has been already studied in [12, 11, 5, 48, 15, 33]. Yet another analysis-based fast algorithm is the butterfly approximation scheme which can be traced back at least to [42] and has found a series of recent applications in [60, 8, 43, 54, 10]. Moreover, it is well known that certain blocks of the discrete Fourier transform are approximately of low rank [38, 57, 13]. Also, the kernel function κ can be approximated by low rank approximations for certain blocks which leads to the butterfly sparse fast Fourier transforms [2, 58]. In this thesis, we follow [58] where the author uses a special interpolation scheme. This scheme factors out the oscillatory part of the restricted Fourier kernel, interpolates the nonoscillatory part, and remodulates the outcome. Based on a local error analysis, we develop a rigorous error analysis for the whole butterfly scheme, which shows how the local expansion degree depends on the target accuracy and the diameter of the frequency set Ω. Moreover, we show that the original scheme becomes numerically unstable if a large local expansion degree is used. We remove this problem by representing all approximations in a Lagrange type basis instead of the previously used monomial type basis. We adduce a complexity analysis of all versions and illustrate the theoretical results by numerical experiments. These findings are published in our paper [35].

In the last part of this thesis, we discuss a fast Fourier transform for nonequispaced complex evaluation nodes in the univariate case. This is already done in [1] as a generalization of the nonequispaced FFT (NFFT). The author of [1] uses convolution techniques and computations by FFTs. Our first idea to compute such sums consists of a generalization of the local butterfly interpolation to complex evaluation nodes. Both approaches allow only the evaluation in complex nodes in a very small strip around the real line. To overcome this issue, we develop another idea which combines both the Fourier and the Laplace transform by means of a multiplicative separation of the oscillatory and the smooth part of the kernel function. We do this in a purely algebraic fashion by decomposing the Laplace transform explicitly and by using a small number of generalized fast Fourier transforms as black box. Alternatively, we might interpret this approach for equispaced frequencies as the fast evaluation of a polynomial given by its monomial coefficients at many nodes in the complex unit disk. The theoretical results on accuracy and computational complexity are illustrated by numerical experiments. Finally, we refer the practically orientated reader to our toolbox [37] which consists of the implementations of the algorithms presented here.

1.1 Outline of the thesis

The thesis is organized as follows.

Chapter 2: Preliminaries and basic definitions.
We introduce most of our notations and give a short overview about tensor decompositions in Section 2.2. Tensors are multi-dimensional arrays and we use the concept [34] to compute the butterfly sparse fast Fourier transform in higher dimensions. Moreover, we summarize in Section 2.3 some facts about polynomial interpolation, in particular the interpolation in Chebyshev nodes which provides a nearly best approximation, see Theorem 2.16. Each local approximation of the kernel functions in our algorithms is based on the interpolation in Chebyshev nodes. In addition, we shortly introduce low rank approximations of matrices and kernel functions. Later, we will use interpolation to create low rank approximations of kernel functions.

Chapter 3: Laplace transform and generalizations.
We consider asymptotically smooth kernels $\kappa\colon Y \times \Omega \to \mathbb{R}$, $Y, \Omega \subset [0,\infty)$, whose q-th partial derivatives have a specified behavior depending on q. Furthermore, a singularity of the kernel κ in $y\xi = 0$ is allowed. Two examples of asymptotically smooth kernels are given by the exponential kernel, which has no singularity, and by the modified Bessel function of the second kind, which has a singularity in $y\xi = 0$, see Lemmata 3.2 and 3.3, respectively. An extension of the results in [46] of the exponential kernel to asymptotically smooth kernels is presented. After introducing the two-dimensional interpolation in a tensor Chebyshev grid, we prove in Theorem 3.5 that the interpolation under an admissible condition leads to a good low rank approximation with an exponentially decreasing error for increasing numbers of interpolation nodes. Section 3.2 contains a summary of the Laplace transform given in [46], where the kernel κ in Equation (1.1) denotes the exponential kernel $\kappa(y,\xi) = \mathrm{e}^{-y\xi}$. We improve the interpolation error [46, Equation (36)] of the whole algorithm by improving the local interpolation error, see Corollary 3.6. Using a geometrical decomposition of the domains Y and Ω gives admissible pairs of subsets. The exponential kernel takes the value one if and only if one of the arguments is zero and the kernel is fast decreasing to zero for large values y or ξ. The algorithm in Theorem 3.9 approximates the kernel function by one, if one subset of Ω or Y of an admissible pair is near the origin and by zero if one of the subsets of Ω or Y is far away from the origin. Between those cases, we approximate the kernel function by interpolation and finally sum up the developed approximations with respect to the coefficients $\hat{f}_k$. In addition, we show that the fast Laplace transform, cf. Algorithm 1, scales linearly in the problem size except logarithmic terms with respect to accuracy. We extend the Laplace transform

to asymptotically smooth kernels in Remark 3.15, where we do exact computations in the near fields and approximations otherwise. We close this chapter with numerical experiments illustrating the theoretical results on accuracy and complexity.

Chapter 4: Butterfly sparse fast Fourier transform.

Now, we aim to compute the sum (1.1) for the kernel $\kappa(\boldsymbol{x}, \boldsymbol{\xi}) = \mathrm{e}^{2\pi\mathrm{i}\boldsymbol{\xi}\boldsymbol{x}/N}$. While the naive computation scales quadratically in the problem size, the FFT for nonequispaced data in spatial and frequency domain [15] or type-3 nonuniform FFT [27] reduces this to a linear complexity in the problem size with an additional $N^d \log N$ term, where N denotes the diameter of Ω or the length of the FFT. Let N^{d-1}-many frequencies $\boldsymbol{\xi}$ and evaluation nodes $\boldsymbol{x}$ be given on smooth $(d-1)$-dimensional manifolds for $d \geq 2$, then the above cited algorithms do not take any advantages of this sparsity constraint on the frequencies and evaluation nodes. The here presented butterfly sparse fast Fourier transform (BSFFT), reduces the complexity of computing the sum (1.1) to a complexity scaling linearly in the problem size except logarithmic terms. We start by introducing the butterfly algorithm in the univariate case. We follow the ideas in [58] and use interpolation in Chebyshev nodes as low rank approximation. In contrast to [58], where the author interpolates the kernel in the spatial and frequency domain in Chebyshev nodes, we interpolate the kernel only in Chebyshev nodes in the spatial domain and use equispaced nodes in the frequency domain. An explicit representation and uniqueness result is stated in Lemma 4.4. In Theorem 4.8, a local error estimate is proven under some admissible condition. An explicit representation of the interpolation problem and an error estimate of the original approach [58] neither contained in [58] nor [35] is declared in Definition 4.12 and Corollary 4.13, respectively. Afterwards, we present the butterfly scheme with the corresponding BSFFT to compute (1.1). We consider two different representations of the interpolating function itself, on the one hand the monomial type basis, which is used in [58], and on the other hand a Lagrange type basis. We discuss the computations of both variants and come up with stability results in Theorem 4.15 and Theorem 4.16.

Section 4.2 presents the butterfly sparse fast Fourier transform in higher dimensions. At first, we extend the interpolation to higher dimensions, declare the algorithm, and discuss the computation of the two realizations of interpolating functions by using tensor decompositions. In Theorem 4.22, we prove an error analysis for the whole algorithm. It is shown that computing the sum (1.1) by the BSFFT reduces the complexity from $\mathcal{O}(N^2)$ to $\mathcal{O}(N \log N (|\log \varepsilon| + \log N)^2)$ floating point operations in the one-dimensional case and from $\mathcal{O}(N^{2(d-1)})$ to $\mathcal{O}(N^{d-1} \log N (|\log \varepsilon| + \log N)^{d+1})$ in the d-dimensional case, where ε describes the target accuracy. Afterwards, we discuss the space complexity of the algorithm and conclude the chapter with numerical experiments illustrating the theoretical results. Finally, a simple application follows, where we treat the vibrating string problem and approximate its solution with a two-dimensional BSFFT.

Chapter 5 : Fast Fourier transform for nonequispaced complex nodes.

In this chapter we consider the Fourier kernel $\kappa(w, \xi) = e^{2\pi i w \xi / N}$ with a complex first argument $w \in \mathbb{C}$. A first idea is the extension of the interpolation operator given by Definition 4.1 to a small complex strip around the real line. We prove a local error estimate in Theorem 5.4 which allows for a certain target accuracy if one chooses a sufficiently small strip depending on the approximation rank. To evaluate the sum (1.1) for arbitrary complex nodes w on the nonnegative half-plane we develop Algorithm 7 which combines the Laplace and Fourier transform by means of a separation of the oscillatory and the smooth part of the kernel function. Algorithm 7 takes at most $\mathcal{O}\left(N \log N \log^2 \frac{N}{\varepsilon} \log \frac{N y_1}{\varepsilon} \log \frac{1}{\varepsilon}\right)$ floating point operations using the nonequispaced FFT in time and frequency (NNFFT) or the BSFFT for the Fourier part, where y_1 denotes the diameter of the real part of the spatial domain. Furthermore, we state an error analysis in Theorem 5.8. Afterwards, we describe how to apply Algorithm 7 for evaluating polynomials in the unit disk. We close this chapter with some numerical experiments. In particular we compare Algorithm 7 with [1, Algorithm 2] for evaluating complex polynomials in the unit disk.

Chapter 6 : Software library.

The Laplace Transform (Algorithm 1), the BSFFT for the dimensions $d = 1, 2, 3, 4$ (Algorithm 3 and 5), and the fast Fourier transform for complex evaluation nodes (Algorithm 7) are implemented in one MATLAB toolbox called extended fast Fourier transforms (XFFT). We give a short tutorial for using this toolbox.

1.2 Acknowledgements

First of all, I wish to express my sincere thanks to my family for supporting me in all situations of my life. I thank my doctoral advisor Prof. Stefan Kunis for the opportunity to realize this thesis. I highly appreciate his valuable advices and ideas throughout the years. I also thank Prof. Laurent Demanet for agreeing to provide a review of this thesis. Moreover, I thank all my colleagues, in particular Christina Brandt, Torsten Görner, Susanne Kunis, Thomas Peter and Ulrich von der Ohe for contributing to an excellent working climate and fruitful on- and off-topic discussions. Finally, I highly appreciate all my amazing friends outside my work for their moral support over the years. Special thanks goes to my physical therapist Eric for his patience and help to get back on track.

Chapter 2

Preliminaries and basic definitions

In this chapter, we introduce some basic notations and definitions for this thesis. Furthermore, an overview in the fundamentals of tensor decompositions, polynomial interpolation, in particular polynomial interpolation in Chebyshev nodes are given. Finally, we introduce low rank approximations.

2.1 Notation

We denote the natural numbers, the natural numbers with zero, the integers, the real numbers, the positive real numbers and the complex numbers by $\mathbb{N}$, $\mathbb{N}_0$, $\mathbb{Z}$, $\mathbb{R}$, $\mathbb{R}^+$, and $\mathbb{C}$, respectively. Throughout this thesis, we set $\mathbb{K} \in \{\mathbb{R}, \mathbb{C}\}$. Furthermore, the spaces of polynomials of degree at most $p \in \mathbb{N}$ are defined by

$$\Pi_p(\mathbb{C}) := \left\{ q \colon \mathbb{C} \to \mathbb{C} : q(z) = \sum_{j=0}^{p} c_j z^j, \ c_j \in \mathbb{C} \right\},$$

$$\Pi_p(\mathbb{R}) := \left\{ q \colon \mathbb{R} \to \mathbb{R} : q(x) = \sum_{j=0}^{p} c_j x^j, \ c_j \in \mathbb{R} \right\},$$

$$\Pi_p(\mathbb{R}, \mathbb{C}) := \left\{ q \colon \mathbb{R} \to \mathbb{C} : q(x) = \sum_{j=0}^{p} c_j x^j, \ c_j \in \mathbb{C} \right\}.$$

Let $d \in \mathbb{N}$ and $D \subset \mathbb{K}^d$ compact. We denote the space of continuous functions $f \colon D \to \mathbb{K}$ by $C(D)$ with the norm $\|f\|_{C(D)} := \max_{x \in D} |f(x)|$. Moreover, we denote the partial derivative of a function $f \colon \mathbb{R}^n \to \mathbb{R}$ in one variable x by $\partial_x f := \frac{\partial}{\partial_x} f$. For $d = 1$, $D \subset \mathbb{R}$, and $m \in \mathbb{N}$, we define the space of all m-times differentiable and continuous functions $f \colon D \to \mathbb{K}$ by $C^{(m)}(D)$. Furthermore, we denote the m-th derivative of f by

$f^{(m)}$. Let $\alpha \in \mathbb{C}$ and $k \in \mathbb{Z}$. The generalized binomial coefficient is defined by

$$
\binom{\alpha}{k} := \begin{cases} \frac{\alpha(\alpha-1)\cdots\cdot(\alpha-(k-1))}{k!} & \text{for } k > 0, \\ 1 & \text{for } k = 0, \\ 0 & \text{for } k < 0. \end{cases}
$$

We call a closed and bounded interval $A = [a,b]$ box and define the diameter by $\operatorname{diam} A := b - a$. Moreover, we define the distance from the interval A to a point $d \in \mathbb{R}$ by $\operatorname{dist}(A,d) := \min_{x \in A} |x - d|$, and the center of the box A by $c^A := \frac{a+b}{2}$. A d-dimensional box is given by $A = A_1 \times \cdots \times A_d \subset \mathbb{R}^d$, where A_i, $i = 1, \ldots, d$, are one-dimensional boxes. The center of the d-dimensional box is given by $c^A := \left(c^{A_1}, \ldots, c^{A_d}\right)$ and the diameter by $\operatorname{diam} A := \max_i \operatorname{diam} A_i$.

Let $\mathbf{v} \in \mathbb{C}^d$. The transpose of the vector $\mathbf{v}$ is denoted by $\mathbf{v}^\top$ and the conjugate transpose by $\mathbf{v}^*$. Moreover, let $a \in \mathbb{C}$, and $\mathbf{e} := (1, \ldots, 1)^\top \in \mathbb{C}^d$, then the shifted vector $\mathbf{v} - a := \mathbf{v} - a\mathbf{e}$ is defined. Furthermore, $\operatorname{diag}(\mathbf{v})$ denotes the diagonal matrix with vector $\mathbf{v}$ on the diagonal. Let $\mathbf{a}, \mathbf{b} \in \mathbb{C}^d$, we define the scalar product $\mathbf{ab} := \mathbf{a}^*\mathbf{b}$ and the norms

$$
\|\mathbf{a}\|_p := \sqrt[p]{\sum_{j=1}^{d} |a_j|^p} \quad \text{and} \quad \|\mathbf{a}\|_\infty := \max_{j=1,\ldots,d} |a_j|
$$

for $p \in \mathbb{N}$, $p \geq 1$. Furthermore, we denote ℓ^p as the sequence space, which includes all sequences $a = (a_n)_{n \in \mathbb{N}}$, such that $\|a\|_p := \left(\sum_{j \in \mathbb{N}} |a_j|^p\right)^{1/p} < \infty$.

Let $\mathbf{A} \in \mathbb{K}^{m \times n}$, $m, n \in \mathbb{N}$, we denote by $\mathbf{A}^\top$ the transposed matrix, by $\mathbf{A}^*$ the adjoint matrix and by $\operatorname{rank}(\mathbf{A})$ the rank of the matrix $\mathbf{A}$. Furthermore, we write $\|\cdot\|_2$ for the spectral norm and $\|\cdot\|_F$ for the Frobenius norm,

$$
\|\mathbf{A}\|_2 := \max_{\mathbf{x} \in \mathbb{K}^n \setminus \{0\}} \frac{\|\mathbf{A}\mathbf{x}\|_2}{\|\mathbf{x}\|_2} \quad \text{and} \quad \|\mathbf{A}\|_F := \sqrt{\sum_{i=1}^{m} \sum_{j=1}^{n} |a_{i,j}|^2}.
$$

Moreover, we define the absolute maximum entry of the matrix $\mathbf{A}$ by

$$
\|\mathbf{A}\|_{1 \to \infty} := \max_{\substack{i=1,\ldots,m \\ j=1,\ldots,n}} |a_{i,j}|, \tag{2.1}
$$

and the Kronecker delta by

$$
\delta_{k,j} := \begin{cases} 1 & \text{for } j = k, \\ 0 & \text{otherwise} \end{cases}
$$

for $k, j \in \mathbb{N}$.

Lemma 2.1. *For $\mathbf{A} \in \mathbb{C}^{m \times n}$, we have*

$$
\|\mathbf{A}\|_{1 \to \infty} = \max_{\substack{\mathbf{f} \in \mathbb{C}^n \\ \|\mathbf{f}\|_1 = 1}} \|\mathbf{A}\mathbf{f}\|_\infty.
$$

Proof. We have

$$\max_{\substack{\mathbf{f}\in\mathbb{C}^n \\ \|\mathbf{f}\|_1=1}} \|\mathbf{Af}\|_\infty \leq \|\mathbf{A}\|_{1\to\infty} \max_{\substack{\mathbf{f}\in\mathbb{C}^n \\ \|\mathbf{f}\|_1=1}} \|\mathbf{f}\|_1 = \|\mathbf{A}\|_{1\to\infty}.$$

Choosing $1 \leq k \leq n$, such that $\|\mathbf{A}\|_{1\to\infty} = \max_{i=1,\ldots,m} |a_{i,k}|$ implies with the k-th unit vector $\mathbf{e}_k := (\delta_{k,j})_{j=1,\ldots,n}$ the assertion,

$$\max_{\substack{\mathbf{f}\in\mathbb{C}^n \\ \|\mathbf{f}\|_1=1}} \|\mathbf{Af}\|_\infty \geq \|\mathbf{Ae}_k\|_\infty = \max_{i=1,\ldots,m} \left| \sum_{j=1}^n a_{i,j}\delta_{k,j} \right| = \max_{i=1,\ldots,m} |a_{i,k}| = \|\mathbf{A}\|_{1\to\infty}.$$

$\square$

For $m = n$ and $\mathbf{A}$ invertible, we denote by $\kappa(\mathbf{A})$ the condition number of $\mathbf{A}$ in the spectral norm.

Furthermore, let $x \in \mathbb{R}$, we define the floor, $\lfloor\cdot\rfloor \colon \mathbb{R} \to \mathbb{Z}$, and ceil function, $\lceil\cdot\rceil \colon \mathbb{R} \to \mathbb{Z}$, by

$$\lfloor x \rfloor := \max\{k \in \mathbb{Z} : k \leq x\},$$
$$\lceil x \rceil := \min\{k \in \mathbb{Z} : k \geq x\}.$$

Let $g \colon \mathbb{R} \to \mathbb{R}$, we use the Landau notation $\mathcal{O}$ and o,

$$\mathcal{O}(g) := \{f \colon \mathbb{R} \to \mathbb{R}^+ : \exists C > 0 \; \exists n_0 \in \mathbb{R} : f(n) \leq Cg(n) \; \forall n \geq n_0\},$$
$$\mathrm{o}(g) := \{f \colon \mathbb{R} \to \mathbb{R}^+ : \forall C > 0 \; \exists n_0 \in \mathbb{R} : f(n) \leq Cg(n) \; \forall n \geq n_0\}.$$

2.2 Tensor decomposition

To expand the Fourier transforms to higher dimensions, we give a short introduction to tensors which are multi-dimensional arrays. For simplicity, we start with some definitions for matrices, which are two-dimensional tensors. Afterwards, we generalize matrices to tensors and introduce some calculation rules.

We use the same notation for tensors as in [34]. The pointwise product of two matrices $\mathbf{A} = (a_{m,n})_{m,n=1}^{M,N}$, $\mathbf{B} = (b_{k,l})_{k,l=1}^{M,N} \in \mathbb{C}^{M\times N}$ is defined by

$$\mathbf{A} \odot \mathbf{B} := (a_{m,n}b_{m,n})_{m,n=1}^{M,N}.$$

Moreover, we denote the Kronecker product for matrices $\mathbf{A} = (a_{m,n})_{m=1,n=1}^{M,N} \in \mathbb{C}^{M\times N}$ and $\mathbf{B} = (b_{k,l})_{k=1,l=1}^{K,L} \in \mathbb{C}^{K\times L}$ in multi index notation by

$$\mathbf{A} \otimes \mathbf{B} := (a_{m,n}\mathbf{B})_{m=1,n=1}^{M,N} = (a_{m,n}b_{k,l})_{m=1,k=1;n=1,l=1}^{M,K;N,L} \in \mathbb{C}^{MK\times NL}.$$

The vectorization of a matrix $\mathbf{A} = (a_{m,n})_{m=1,n=1}^{M,N}$ is denoted by vec$\colon \mathbb{C}^{M\times N} \to \mathbb{C}^{MN}$,

$$\mathrm{vec}(\mathbf{A}) := (a_{m,n})_{n=1,\ldots,N;m=1,\ldots,N} = (a_{1,1}\ldots a_{m,1}a_{1,2}\ldots a_{m,2}\ldots a_{1,n}\ldots a_{m,n}) \in \mathbb{C}^{MN}$$

which means, that the row index runs first. The vectorization of a matrix is writing the columns of the matrix one below the other.

Example 2.2. *Let* $\mathbf{A} = \begin{pmatrix} 1 & 2 & 3 & 4 \\ 5 & 6 & 7 & 8 \end{pmatrix}$, *then* $\operatorname{vec}(\mathbf{A}) = \begin{pmatrix} 1 & 5 & 2 & 6 & 3 & 7 & 4 & 8 \end{pmatrix}^{\top}$.

Lemma 2.3. *(i) Let* $\mathbf{A} \in \mathbb{C}^{M \times N}$, $\mathbf{B} \in \mathbb{C}^{K \times L}$, $\mathbf{C} \in \mathbb{C}^{N \times J}$, *and* $\mathbf{D} \in \mathbb{C}^{L \times G}$ *be given, then it follows*

$$\mathbf{AC} \otimes \mathbf{BD} = (\mathbf{A} \otimes \mathbf{B})(\mathbf{C} \otimes \mathbf{D}).$$

(ii) Two invertible matrices $\mathbf{A} \in \mathbb{C}^{N \times N}$, $\mathbf{B} \in \mathbb{C}^{M \times M}$, *satisfy*

$$(\mathbf{A} \otimes \mathbf{B})^{-1} = \mathbf{A}^{-1} \otimes \mathbf{B}^{-1}. \tag{2.2}$$

(iii) For $\mathbf{A} \in \mathbb{C}^{M \times N}$, $\mathbf{B} \in \mathbb{C}^{K \times L}$ *and* $\mathbf{U} \in \mathbb{C}^{L \times N}$, *we have*

$$\operatorname{vec}(\mathbf{F}) = (\mathbf{A} \otimes \mathbf{B}) \operatorname{vec}(\mathbf{U}) \Leftrightarrow \mathbf{F} = \mathbf{BUA}^{\top}.$$

(iv) Let $\mathbf{v}, \mathbf{w} \in \mathbb{C}^N$ *and a matrix* $\mathbf{U} \in \mathbb{C}^{N \times N}$ *be given, then it follows*

$$(\operatorname{diag} \mathbf{v} \otimes \operatorname{diag} \mathbf{w}) \operatorname{vec}(\mathbf{U}) = \operatorname{vec}\left(\mathbf{w}\mathbf{v}^{\top} \odot \mathbf{U}\right).$$

Proof. See [47, 1.3 Matrix Multiplication] for (i) and (ii) and [47, 1.12 Vec operator, Theorem] for (iii). We prove (iv). From (iii), we conclude

$$(\operatorname{diag}(\mathbf{v}) \otimes \operatorname{diag}(\mathbf{w})) \operatorname{vec}(\mathbf{U}) = \operatorname{vec}(\operatorname{diag}(\mathbf{w})\mathbf{U}\operatorname{diag}(\mathbf{v}))$$

and hence the assertion follows since

$$(\operatorname{diag}(\mathbf{w})\mathbf{U}\operatorname{diag}(\mathbf{v}))_{i,j} = w_i u_{i,j} v_j = \left((\mathbf{w}\mathbf{v}^{\top}) \odot \mathbf{U}\right)_{i,j}.$$

$\square$

We generalize the Kronecker product to higher dimensions using tensors. At first, we use the notation of a multi index vector $\mathbf{f} \in \mathbb{C}^{N_1 \cdots N_d}$ given by

$$\mathbf{f} = (f_{n_1,\dots,n_d})_{n_1=1,\dots,N_1,\dots,n_d=1,\dots,N_d} \in \mathbb{C}^{N_1 \cdots N_d},$$

where the last index n_d runs first and then the previous one, n_{d-1}, and so on up to n_1. By abuse of notation, we write the usual Kronecker product for matrices $\mathbf{A}^{(j)} \in \mathbb{C}^{M_j \times N_j}$, $M_j, N_j \in \mathbb{N}$, and $j = 1, \dots, d$, in a multi index notation

$$\bigotimes_{\nu=1}^{d} \mathbf{A}^{(j)} = \left(\prod_{\nu=1}^{d} \mathbf{A}_{m_\nu,n_\nu}^{(\nu)}\right)_{m_1=1,\dots,m_d=1;n_1=1,\dots,n_d=1}^{M_1,\dots,M_d;N_1,\dots,N_d} \in \mathbb{C}^{M_1 \cdots M_d \times N_1 \cdots N_d}.$$

For a fixed multi row index $m_1, \dots, m_d$ the column index n_d runs first from 1 to N_d and afterwards n_{d-1} from 1 to N_{d-1} and so on up to n_1. The multi row indices are traversed in the same way. The number of dimensions of a tensor is called order of a tensor, also known as ways or modes. The next definition is a generalization of the vectorization of matrices to tensors.

Definition 2.4. *Let* $d \in \mathbb{N}$, $d \geq 2$, *and* $N_1, \ldots, N_d \in \mathbb{N}$. *We denote a d-tensor of dimension* $N_1 \times \cdots \times N_d$ *by* $\mathcal{F} = (f_{n_1,\ldots,n_d})_{n_1=1,\ldots,n_d=1}^{N_1,\ldots,N_d}$. *The numbers* $f_{n_1,\ldots,n_d} \in \mathbb{C}$ *are called* tensor entries. *The* vec-*operator is defined by*

$$\text{vec} \colon \mathbb{C}^{N_1 \times \cdots \times N_d} \to \mathbb{C}^{N_1 \cdots N_d}, \quad \text{vec}(\mathcal{F}) := (f_{n_1,\ldots,n_d})_{n_d=1,\ldots,N_d,\ldots,n_1=1,\ldots,N_1},$$

where the index n_1 *runs first and then* n_2, *and so on up to* n_d.

From now on, we consider for simplicity only tensors where every mode is the same size. Such tensors are called cubical tensors. We only need cubical tensors in this thesis.

Definition 2.5. *The pointwise product of two d-tensors* $\mathcal{F}$ *and* $\mathcal{U} = (u_{n_1,\ldots,n_d})_{n_1,\ldots,n_d=1}^{N}$ *is defined by*

$$\mathcal{F} \odot \mathcal{U} := (f_{n_1,\ldots,n_d} \cdot u_{n_1,\ldots,n_d})_{n_1,\ldots,n_d=1}^{N}.$$

Let $k \in \mathbb{N}$ *and* $1 \leq k \leq d$. *The* k-mode product *of a tensor* $\mathcal{F}$ *and a matrix* $\mathbf{A} := (a_{m,n_k})_{m=1,n_k=1}^{M,N} \in \mathbb{C}^{M \times N}$, $M \in \mathbb{N}$, *is defined by* $\mathcal{U} \times_k \mathbf{A} \in \mathbb{C}^{N \times \cdots \times N \times M \times N \times \cdots \times N}$,

$$(\mathcal{U} \times_k \mathbf{A})_{n_1 \ldots n_{k-1} m n_{k+1} \ldots n_d} := \sum_{n_k=1}^{N} u_{n_1 \ldots n_d} a_{mn_k}.$$

Furthermore, let $1 \leq k \leq d$. *The* k-mode matricization *of a d-tensor* $\mathcal{F} \in \mathbb{C}^{N \times \cdots \times N}$ *is denoted by* $\mathbf{F}_{(k)} \in \mathbb{C}^{N \times N^{d-1}}$, *where the tensor entry* $(n_1, \ldots, n_d)$ *is mapped to the matrix entry* (n_k, m) *with* $m = 1 + \sum\limits_{\substack{j=1 \\ j \neq k}}^{d} (n_j - 1) A_j$, $A_j := \begin{cases} N^{j-2} & \text{for } j \geq k+1 \\ N^{j-1} & \text{for } j < k. \end{cases}$

Example 2.6 ([34]). (i) *Let a matrix (a 2-tensor)* $\mathbf{A} = (a_{ij})_{i=1,j=1}^{M,N}$ *be given. The mode n-foldings are*

$$\mathbf{A}_{(1)} = \mathbf{A} \quad \text{and} \quad \mathbf{A}_{(2)} = \mathbf{A}^{\top}.$$

The vectorization coincides with the definition of the vectorization of a matrix.

(ii) *Let* $\mathcal{U} \in \mathbb{C}^{2 \times 2 \times 2}$, *where* $u_{i,j,k} \in \mathbb{C}$ *for* $i, j, k = 1, 2$ *given by*

$$\mathcal{U}(:,:,1) = \begin{pmatrix} u_{1,1,1} & u_{1,2,1} \\ u_{2,1,1} & u_{2,2,1} \end{pmatrix}, \quad \mathcal{U}(:,:,2) = \begin{pmatrix} u_{1,1,2} & u_{1,2,2} \\ u_{2,1,2} & u_{2,2,2} \end{pmatrix}.$$

The three mode-n foldings are $\mathbf{U}_{(j)} \in \mathbb{R}^{2 \times 4}$ *for* $j = 1, 2, 3$,

$$\mathbf{U}_{(1)} = \begin{pmatrix} u_{1,1,1} & u_{1,2,1} & u_{1,1,2} & u_{1,2,2} \\ u_{2,1,1} & u_{2,2,1} & u_{2,1,2} & u_{2,2,2} \end{pmatrix},$$

$$\mathbf{U}_{(2)} = \begin{pmatrix} u_{1,1,1} & u_{2,1,1} & u_{1,1,2} & u_{2,1,2} \\ u_{1,2,1} & u_{2,2,1} & u_{1,2,2} & u_{2,2,2} \end{pmatrix},$$

$$\mathbf{U}_{(3)} = \begin{pmatrix} u_{1,1,1} & u_{2,1,1} & u_{1,2,1} & u_{2,2,1} \\ u_{1,1,2} & u_{2,1,2} & u_{1,2,2} & u_{2,2,2} \end{pmatrix}.$$

The vectorization of $\mathcal{U}$ *is given by*

$$\text{vec}\,\mathcal{U} = (u_{1,1,1}, u_{2,1,1}, u_{1,2,1}, u_{2,2,1}, u_{1,1,2}, u_{2,1,2}, u_{1,2,2}, u_{2,2,2})^{\top}.$$

Lemma 2.7 ([34]). *Let* $N, d \in \mathbb{N}$ *and* $d \geq 2$, *a* d-*tensor* $\mathcal{U} \in \mathbb{C}^{N \times \cdots \times N}$, *and two matrices* $\mathbf{A} := (a_{l,m})_{l,m=1}^N \in \mathbb{C}^{N \times N}$, $\mathbf{B} := (b_{l,m})_{l,m=1}^N \in \mathbb{C}^{N \times N}$ *be given.*

(i) *Let* $\mathbf{I}_k := (\delta_{i,j})_{i,j=1}^{N^k}$, $k \in \mathbb{N}_0$, *be the identity, then it follows*

$$(\mathbf{I}_{k-1} \otimes \mathbf{A} \otimes \mathbf{I}_{d-k}) \operatorname{vec} \mathcal{U} = \operatorname{vec}(\mathcal{U} \times_{d-k+1} \mathbf{A}) \quad \text{for } k \in \mathbb{N}, \ 1 \leq k \leq d.$$

(ii) *The* k-*mode product of* $\mathcal{U}$ *and* $\mathbf{A}$ *satisfies*

$$\mathcal{Y} = \mathcal{U} \times_k \mathbf{A} \Leftrightarrow \mathbf{Y}_{(k)} = \mathbf{A} \mathbf{U}_{(k)} \quad \text{for } k \in \mathbb{N}, \ 1 \leq k \leq d.$$

(iii) *Let* $\mathbf{v}^{(1)}, \ldots, \mathbf{v}^{(d)} \in \mathbb{C}^N$ *with* $\mathbf{v}^{(i)} := (v_{n_i})_{n_i=1}^N$. *Then it follows*

$$\left(\operatorname{diag}(\mathbf{v}^{(1)}) \otimes \cdots \otimes \operatorname{diag}(\mathbf{v}^{(d)}) \right) \operatorname{vec}(\mathcal{U}) = \operatorname{vec}(\mathcal{V} \odot \mathcal{U}),$$

where $\mathcal{V} := \left(v_{n_1}^{(1)} \cdots v_{n_d}^{(d)} \right)_{n_1,\ldots,n_d=1}^N \in \mathbb{C}^{N \times \cdots \times N}$.

(iv) *Let* $d = 2$ *and* $\mathcal{U} = \mathbf{U} \in \mathbb{C}^{N \times N}$. *The calculation rule of Kronecker products agrees with the tensor calculation rules,*

$$(\mathbf{A} \otimes \mathbf{B}) \operatorname{vec}(\mathbf{U}) = \operatorname{vec}(\mathbf{U} \times_1 \mathbf{B} \times_2 \mathbf{A}) = \operatorname{vec}(\mathbf{B} \mathbf{U} \mathbf{A}^\top).$$

Proof. (i) The identities $\mathbf{I}_k \in \mathbb{C}^{N^k \times N^k}$, $k \in \mathbb{N}$, can be written as k Kronecker products of the identity $\mathbf{I}_1$, $\mathbf{I}_k = \bigotimes_{j=1}^k \mathbf{I}_1 = (\delta_{r_1,t_1} \cdots \delta_{r_k,t_k})_{r_1,\ldots,r_k;t_1,\ldots,t_k=1}^N$. We obtain

$$
\begin{aligned}
\mathbf{I}_{k-1} \otimes \mathbf{A} \otimes \mathbf{I}_{d-k} &= \left(\delta_{r_1,t_1} \cdots \delta_{r_{k-1},t_{k-1}} \mathbf{A} \right)_{r_1,\ldots,r_{k-1}=1;t_1,\ldots,t_{k-1}=1}^{N;N} \otimes \mathbf{I}_{d-k} \\
&= \left(\delta_{r_1,t_1} \cdots \delta_{r_{k-1},t_{k-1}} a_{l,m} \right)_{r_1,\ldots,r_{k-1},l=1;t_1,\ldots,t_{k-1},m=1}^{N;N} \otimes \mathbf{I}_{d-k} \\
&= \left(\delta_{r_1,t_1} \cdots \delta_{r_{k-1},t_{k-1}} a_{t_k,r_k} \delta_{r_{k+1},t_{k+1}} \cdots \delta_{r_d,t_d} \right)_{r_1,\ldots,r_d=1;t_1,\ldots,t_d=1}^{N;N}.
\end{aligned}
$$

Since $\operatorname{vec}(\mathcal{U}) = (u_{t_1},\ldots,t_d)_{t_d,\ldots,t_1=1,\ldots,N}$, we change the indices $k \mapsto d-k+1$ and obtain

$$\mathbf{I}_{k-1} \otimes \mathbf{A} \otimes \mathbf{I}_{d-k} = (a_{t_{d-k+1},r_{d-k+1}} \prod_{\substack{j=1 \\ j \neq d-k+1}}^{d} \delta_{r_j,l_j})_{r_d,\ldots,r_1=1;t_d,\ldots,t_1=1}^{N;N}$$

Setting $\mathbf{f} := (f_{r_1,\ldots,r_d})_{r_d,\ldots,r_1=1,\ldots,N} = (\mathbf{I}_{k-1} \otimes \mathbf{A} \otimes \mathbf{I}_{d-k}) \operatorname{vec} \mathcal{U} \in \mathbb{C}^{N^d}$, we have

$$
\begin{aligned}
f_{r_1,\ldots,r_d} &= \sum_{t_1,\ldots,t_d=1}^N \delta_{r_d,t_d} \cdots \delta_{r_{d-k+2},t_{d-k+2}} a_{t_{d-k+1},r_{d-k+1}} \delta_{r_{d-k},t_{d-k}} \cdots \delta_{r_1,t_1} u_{t_1,\ldots,t_d} \\
&= \sum_{t_{d-k+1}=1}^N a_{t_{d-k+1},r_{d-k+1}} u_{r_1,\ldots,r_d} = (\mathcal{U} \times_{d-k+1} \mathbf{A})_{r_1,\ldots,r_d}.
\end{aligned}
$$

(ii) Cf. [34, 2.5 Tensor Multiplication: The n-Mode Product].

(iii) The assertion follows since $\mathrm{diag}(\mathbf{v}^{(1)}) \otimes \cdots \otimes \mathrm{diag}(\mathbf{v}^{(n)}) = \mathrm{diag}(\mathbf{v}^{(1)} \otimes \cdots \otimes \mathbf{v}^{(d)})$ and $\mathrm{vec}\,\mathcal{V} = \mathbf{v}^{(1)} \otimes \cdots \otimes \mathbf{v}^{(d)}$.

(iv) We prove that

$$\mathbf{U} \times_1 \mathbf{B} \times_2 \mathbf{A} = \mathbf{BUA}^\top$$

holds true. We set $\mathbf{Z} = \mathbf{Y} \times_2 \mathbf{A}$, where $\mathbf{Y} = \mathbf{U} \times_1 \mathbf{B}$. Applying (ii) and using $\mathbf{Y}_{(1)} = \mathbf{Y}$ and $\mathbf{Z}_{(2)} = \mathbf{Z}^\top$ imply

$$\mathbf{Z}^\top = \mathbf{AY}^\top, \quad \mathbf{Y} = \mathbf{BU},$$

and finally

$$\mathbf{Z} = \mathbf{YA}^\top = \mathbf{BUA}^\top.$$

Due to Lemma (iii), the assertion follows.

$\square$

2.3 Polynomial interpolation

The following problem is called interpolation problem. We have given a set of $p \in \mathbb{N}$ pairwise distinct points $\boldsymbol{x} = (x_s)_{s=0,\ldots,p-1} \in \mathbb{K}^p$ as well as a vector of function values $\mathbf{f} = (f(x_s))_{s=0,\ldots,p-1} \in \mathbb{K}^p$. The task is finding a polynomial $r \in \Pi_{p-1}(\mathbb{K})$ which satisfies the interpolation condition

$$r(x_s) = f(x_s) \text{ for all } s = 0, \ldots, p-1.$$

Definition 2.8. *We denote the nodal polynomial to the nodes $\boldsymbol{x}$ by $w_{\boldsymbol{x}} \in \Pi_p(\mathbb{K})$,*

$$w_{\boldsymbol{x}}(x) := \prod_{s=0}^{p-1}(x - x_s),$$

and the Lagrange polynomials to the nodes $\boldsymbol{x}$ by $L_{\boldsymbol{x},k} \in \Pi_{p-1}(\mathbb{K})$, $k = 0, \ldots, p-1$,

$$L_{\boldsymbol{x},k}(x) := \frac{w(x)}{(x - x_k)w'(x_k)} = \prod_{\substack{s=0 \\ s \neq k}}^{p-1} \frac{x - x_s}{x_k - x_s}. \tag{2.3}$$

Polynomial interpolation is unique.

Theorem 2.9 ([31, Satz 37.3 and 37.4.]). *Let $f \colon \mathbb{K} \to \mathbb{K}$. Under the above assumptions, the interpolation problem has a unique solution $r \in \Pi_{p-1}(\mathbb{K})$ given by*

$$r(x) = \sum_{s=0}^{p-1} f(x_s)L_{\boldsymbol{x},s}(x).$$

Moreover, let now $\mathbb{K} = \mathbb{R}$ and $f \colon [x_0, x_{p-1}] \to \mathbb{R}$, $f \in C^p[x_0, x_{p-1}]$. For $x \in [x_0, x_{p-1}]$ exists $\xi(x) \in (x_0, x_{p-1})$, such that the interpolation error is given by

$$f(x) - r(x) = \frac{f^{(p)}(\xi(x))}{p!} w(x).$$

2.3.1 Interpolation in Chebyshev nodes

We introduce Chebyshev polynomials, which are orthogonal polynomials and very important in numerical analysis, see [17, 45, 40, 53]. Let $p \in \mathbb{N}$. Interpolation in the roots of Chebyshev polynomials provides a nearly best approximation as stated in [53, Theorem 16.1]. The Chebyshev polynomials of the first kind are defined by $T_p \colon \mathbb{R} \to \mathbb{R}$,

$$T_p(x) := \begin{cases} \cos(p \arccos x) & \text{for } x \in [-1, 1], \\ \frac{(x+\sqrt{x^2-1})^p + (x-\sqrt{x^2-1})^p}{2} & \text{otherwise.} \end{cases}$$

The Chebyshev polynomials obey the following three-term recurrence relation

$$T_{p+1}(x) = 2xT_p(x) - T_{p-1}(x), \quad T_0(x) = 1, \quad T_1(x) = x. \tag{2.4}$$

It is obvious, that we obtain $T_p \in \Pi_p(\mathbb{R})$. We denote the roots of the Chebyshev polynomial T_p by

$$\boldsymbol{t} := (t_j)_{j=0,\dots,p-1}, \quad t_j := \cos \frac{2j+1}{2p} \pi,$$

and we call them Chebyshev nodes. Furthermore, the restricted Chebyshev polynomial $T_p \colon [-1, 1] \to \mathbb{R}$ takes its extrema at the nodes

$$t_j^{\max} := \cos \frac{j\pi}{p}, \quad j = 0, \dots, p.$$

Moreover, we introduce the Chebyshev polynomials of the second kind by $U_p \colon \mathbb{R} \to \mathbb{R}$,

$$U_p(x) := \begin{cases} \frac{\sin((p+1)\arccos x)}{\sin(\arccos x)} & \text{for } x \in (-1, 1), \\ (\pm 1)^p (p+1) & \text{for } x = \pm 1, \\ \frac{(x+\sqrt{x^2-1})^{p+1} - (x-\sqrt{x^2-1})^{p+1}}{2\sqrt{x^2-1}} & \text{otherwise,} \end{cases}$$

which obey the similar three-term recurrence relation (2.4) with different initial conditions,

$$U_{p+1}(x) = 2xU_p(x) - U_{p-1}(x), \quad U_0(x) = 1, \quad U_1(x) = 2x.$$

Lemma 2.10. *Let A be a box and $p \in \mathbb{N}$. We call*

$$\boldsymbol{x}^A := (c^A + \frac{\operatorname{diam} A}{2} t_j)_{j=0,\dots,p-1} \tag{2.5}$$

the Chebyshev nodes in the box A. The Chebyshev polynomials, the Chebyshev nodes and the corresponding nodal polynomial obey

$$T_p'(x) = pU_{p-1}(x), \tag{2.6}$$

$$\|w_{\boldsymbol{t}}\|_{C[-1,1]} = 2^{-(p-1)}, \tag{2.7}$$

$$\|w_{\boldsymbol{x}^A}\|_{C(A)} = 2 \left(\frac{\operatorname{diam} A}{4} \right)^p, \quad and \tag{2.8}$$

$$\prod_{\substack{j=0 \\ j \neq r}}^{p-1} |t_j| = \frac{p}{2^{p-1}} \quad if \ p = 2r+1, \ r \in \mathbb{N}. \tag{2.9}$$

Proof. The first equation can be found in [45, (1.23)] and the second in [45, Theorem 2.1]. The third equation follows from the second by mapping the interval $[-1, 1] \rightarrow A$ via $t \mapsto c^A - \frac{\operatorname{diam} A}{2} t =: x$, which implies the assertion since

$$w_{\boldsymbol{x}^A}(x) = \prod_{j=0}^{p-1} (x - x_k^A) = \prod_{j=0}^{p-1} -\frac{\operatorname{diam} A}{2}(t - t_k) = \left(-\frac{\operatorname{diam} A}{2}\right)^p w_{\boldsymbol{t}}(t).$$

Let $p = 2r + 1$, $r \in \mathbb{N}$, it follows $t_r = 0$ and $L_r(0) = 1$. Moreover, we have $T_p'(0) = pU_{p-1}(0) = p$, where $U_{p-1}(\cos\theta) = \frac{\sin p\theta}{\sin\theta}$ denotes the Chebyshev polynomial of second kind. Combining $T_p(x) = 2^{p-1} w_{\boldsymbol{t}}(x)$ with Equation (2.3), we conclude

$$L_r(x) = \frac{T_p(x)}{(x - t_r)T_p'(t_r)}. \tag{2.10}$$

The assertion follows from $x = 0$. $\qquad\qquad\square$

Remark 2.11. *The nodal polynomial $w_{\boldsymbol{t}}$ to the Chebyshev nodes is optimal in the sense, that for all polynomials $r \in \Pi_p(\mathbb{R})$ with leading coefficient 1, we have*

$$\|r\|_{C[-1,1]} \geq \|2^{-(p-1)}T_p\|_{C[-1,1]} = 2^{-(p-1)}$$

with equality only if $r = 2^{-(p-1)}T_p$. See [45, Theorem 2.1] for a proof.

For simplicity we denote the Lagrange polynomials for the Chebyshev nodes $\boldsymbol{t}$ and $\boldsymbol{x}^A$ by

$$L_s := L_{\boldsymbol{t},s} \quad \text{and} \quad L_s^A := L_{\boldsymbol{x}^A,s} \quad \text{for } s = 0, \ldots, p - 1. \tag{2.11}$$

Definition 2.12. *Let A be a box and $f\colon A \to \mathbb{K}$, $f \in C(A)$, $p \in \mathbb{N}$, $p \geq 2$. We denote the interpolation operator for Chebyshev nodes $\boldsymbol{x}^A$ in A by*

$$\mathcal{I}_p^A \colon C(A) \to \Pi_{p-1}(\mathbb{R}, \mathbb{K}), \quad \mathcal{I}_p^A f = \sum_{s=0}^{p-1} f(x_s^A) L_s^A.$$

Theorem 2.13. *Let a box A, $f\colon A \to \mathbb{R}$, $f \in C^p(A)$, and $p \in \mathbb{N}$, $p \geq 2$, be given. The interpolation error is bounded by*

$$\|\mathcal{I}_p^A f - f\|_{C(A)} \leq 2\frac{(\operatorname{diam} A)^p}{4^p p!}\|f^{(p)}\|_{C(A)}.$$

Proof. The polynomial interpolation error is given with some $\xi(x) \in (x_{p-1}^A, x_0^A)$ by

$$|(\mathcal{I}_p^A f - f)(x)| = \frac{1}{p!}|f^{(p)}(\xi)||w_{\boldsymbol{x}^A}(x)|,$$

see Theorem 2.9. Applying the maximum norm $\|\cdot\|_{C(A)}$ and using Inequality (2.8) yields the assertion. $\qquad\qquad\square$

Lemma 2.14. *For $p \in \mathbb{N}$, $p \geq 2$, the Lagrange polynomials (2.11) obey*

$$\max_{x \in [-1,1]} \sum_{s=0}^{p-1} (L_s(x))^2 \leq 2, \tag{2.12}$$

$$\max_{x \in [-1,1]} \sum_{s=0}^{p-1} |L_s(x)| \leq 1 + \frac{2}{\pi} \log p, \tag{2.13}$$

$$\max_{x \in [-3,3]} |L_s(x))| \leq \frac{34^{\frac{p}{2}}}{4p}. \tag{2.14}$$

Proof. The Gauss-Chebyshev quadrature yields the discrete orthogonality

$$\sum_{s=0}^{p-1} T_k(t_s) T_l(t_s) = \begin{cases} 0 & \text{for } k \neq l, \\ p & \text{for } k = l = 0, \\ \frac{p}{2} & \text{for } k = l \neq 0, \end{cases}$$

for $k, l = 0, \ldots, p - 1$, see [45, Equation (1.141)], and since $L_s(t_j) = \delta_{s,j}$ also the expansion of the Lagrange polynomials

$$L_s(x) = \frac{2}{p} \sum_{j=0}^{p-1}{}' T_j(t_s) T_j(x) := \frac{1}{p} T_0(t_s) T_0(x) + \frac{2}{p} \sum_{j=1}^{p-1} T_j(t_s) T_j(x),$$

where the prime indicates that the first summand is weighted by $\frac{1}{2}$. Hence, we have

$$\sum_{s=0}^{p-1} (L_s(x))^2 = \sum_{s=0}^{p-1} \frac{4}{p^2} \left(\sum_{k=0}^{p-1}{}' T_k(t_s) T_k(x) \right) \left(\sum_{l=0}^{p-1}{}' T_l(t_s) T_l(x) \right)$$

$$= \frac{4}{p^2} \sum_{k=0}^{p-1}{}' \sum_{l=0}^{p-1}{}' T_k(x) T_l(x) \sum_{s=0}^{p-1} T_k(t_s) T_l(t_s)$$

and since $|T_k(x)| \leq 1$ for $x \in [-1, 1]$ the first claim by

$$\leq \frac{4}{p^2} \left(\frac{1}{4} p + (p - 1) \frac{p}{2} \right) \leq 2.$$

The second estimate is the classical Lebesgue constant, cf. [45, Theorem 1.2]. Finally, we note that Inequality (2.12) implies $|L_s(x))| \leq \sqrt{2}$ for $|x| \leq 1$ and it remains to show the bound for $|x| \in [1, 3]$. Since L_s is a polynomial and has all its zeros inside $[-1, 1]$, it attains its extrema at $x = \pm 3$. Combining Equation (2.10) with (2.6) we obtain

$$L_s(x) = \frac{T_p(x)}{(x - t_s) p U_{p-1}(t_s)}.$$

The explicit formula

$$T_p(x) = \frac{(x + \sqrt{x^2 - 1})^p + (x - \sqrt{x^2 - 1})^p}{2}, \quad |x| \geq 1,$$

implies $|T_p(\pm 3)| \leq 34^{\frac{p}{2}}$. Applying the simple bound $(\pm 3 - t_s)^2 \geq 4$, and

$$(U_{p-1}(t_s))^2 = \frac{\sin^2\left(\frac{2s+1}{2}\pi\right)}{\sin^2\left(\frac{2s+1}{2p}\pi\right)} \geq \sin^2\left(\frac{2s+1}{2}\pi\right) = (-1)^{2s} = 1, \quad s = 0, \ldots, p-1,$$

finally yields the assertion. $\square$

Corollary 2.15 (Lebesgue constant). *Let $p \in \mathbb{N}$, $p \geq 2$, and a box A be given. Then the operator norm is bounded by the so called Lebesgue constant*

$$\|\mathcal{I}_p^A\| := \max_{\substack{f \in C(A) \\ \|f\|_{C(A)} = 1}} \|\mathcal{I}_p^A f\|_{C(A)} = \|\mathcal{I}_p^{[-1,1]}\| \leq 1 + \frac{2}{\pi} \log p.$$

Proof. Let the mapping $A \to [-1,1]$ via $x \mapsto t := \frac{2}{\operatorname{diam} A}(x - c^A)$ be given. Since a function $f \colon A \to \mathbb{C}$, can be expressed by the function $g \colon [-1,1] \to \mathbb{C}$, $g(t) := f(x(t))$, and since we have $L_s^A(x) = L_s(t)$, the norm of the interpolation operator can be written as

$$\|\mathcal{I}_p^A\| := \max_{\substack{f \in C(A) \\ \|f\|_{C(A)} = 1}} \|\mathcal{I}_p^A f\|_{C(A)} = \max_{\substack{f \in C(A) \\ \|f\|_{C(A)} = 1}} \max_{x \in A} \left| \sum_{s=0}^{p-1} L_s^A(x) f(x_s^A) \right|$$

$$= \max_{\substack{g \in C[-1,1] \\ \|g\|_{C[-1,1]} = 1}} \max_{t \in [-1,1]} \left| \sum_{s=0}^{p-1} L_s(t) g(t_s) \right| = \|\mathcal{I}_p^{[-1,1]}\|.$$

Using the triangle inequality and estimate the function values $|g(t_s)|$ by the maximum norm of g, we obtain

$$\|\mathcal{I}_p^{[-1,1]}\| \leq \max_{t \in [-1,1]} \sum_{s=0}^{p-1} |L_s(t)|$$

and thus the assertion follows with Inequality (2.13). $\square$

The next theorem states, that interpolation in Chebyshev nodes is nearly as good as best approximation.

Theorem 2.16 ([53, Theorem 16.1]). *Let $f \in C[-1,1]$ and $p \in \mathbb{N}$. We denote by $q_p^* \in \Pi_{p-1}(\mathbb{R})$ the best approximation of f in the sense that*

$$\|f - q_p^*\|_{C[-1,1]} = \min_{q_p \in \Pi_{p-1}(\mathbb{R})} \|f - q_p\|_{C[-1,1]}.$$

Then, it follows

$$\|f - \mathcal{I}_p^{[-1,1]} f\|_{C[-1,1]} \leq \left(2 + \frac{2}{\pi} \log p\right) \|f - q_p^*\|_{C[-1,1]}.$$

This theorem can be extended to all boxes A instead of $[-1,1]$.

2.4 Low rank approximation

A matrix $\mathbf{A} \in \mathbb{C}^{M \times N}$ with $\text{rank}(\mathbf{A}) = r \leq \min\{M, N\}$ can be represented with $\mathbf{a}_k \in \mathbb{C}^M$, $\mathbf{b}_k \in \mathbb{C}^N$ for $k = 1, \ldots, r$ by

$$\mathbf{A} = \sum_{k=1}^{r} \mathbf{a}_k \mathbf{b}_k^*.$$

Note, that the memory requirements are $r(M + N)$ instead of MN. Furthermore, the matrix-vector multiplication

$$\mathbf{A}\mathbf{x} = \sum_{k=1}^{r} \mathbf{a}_k \mathbf{b}_k^* \mathbf{x} = \sum_{k=1}^{r} \mathbf{a}_k (\mathbf{b}_k^* \mathbf{x})$$

takes $r(M + N)$ operations instead of MN floating point operations. Summarizing, low rank matrices can be stored efficiently and the matrix can be applied efficiently. That is why we are interested in the approximation of matrices by low rank matrices. The most popular low rank approximation of a matrix $\mathbf{K} \in \mathbb{C}^{M \times N}$ is given by the truncated singular value decomposition. Furthermore, it is the best approximation for the spectral and the Frobenius norm in the following sense.

Theorem 2.17 ([31, Definition und Satz 12.1.]). *Let $\mathbf{K} \in \mathbb{C}^{M \times N}$ of rank $k > 0$. Then there exists two unitary matrices $\mathbf{U} \in \mathbb{C}^{M \times M}$, $\mathbf{V} \in \mathbb{C}^{N \times N}$, and a diagonal matrix $\Sigma = \begin{pmatrix} \Sigma_k & 0 \\ 0 & 0 \end{pmatrix} \in \mathbb{C}^{M \times N}$, $\Sigma_k = \text{diag}(\sigma_1, \ldots, \sigma_k)$, where $\sigma_1 \geq \sigma_2 \cdots \geq \sigma_{k-1} \geq \sigma_k > 0$ are called singular values, such that $\mathbf{K} = \mathbf{U} \Sigma \mathbf{V}^*$.*

Theorem 2.18 ([29, Satz 2.4.1], Best rank k approximation). *With the same assumptions as in the previous theorem, the best rank r approximation of $\mathbf{K}$ with respect to $\| \cdot \|_2$ and $\| \cdot \|_F$, for $r \leq k$ is given by $\mathbf{B} = \sum_{i=1}^{r} \mathbf{u}_i \sigma_i \mathbf{v}_i^*$, where we denote the columns of the matrices $\mathbf{U}, \mathbf{V}$ by $\mathbf{u}_i$ and $\mathbf{v}_j$, respectively. Furthermore, it follows*

$$\|\mathbf{K} - \mathbf{B}\|_2 = \sigma_{r+1} \quad and \quad \|\mathbf{K} - \mathbf{B}\|_F = \sqrt{\sum_{i=r+1}^{k} \sigma_i^2}. \tag{2.15}$$

Let $X, Y \subset \mathbb{K}$, $k \in \mathbb{N}$ and functions $\varphi_\nu \colon X \to \mathbb{C}$, $\psi_\nu \colon Y \to \mathbb{C}$, $\nu = 1, \ldots, k$, be given. We call $\tilde{\kappa} \colon X \times Y \to \mathbb{C}$,

$$\tilde{\kappa}(x, y) = \sum_{\nu=1}^{k} \varphi_\nu(x) \psi_\nu(y)$$

a low rank approximation with rank k to the function κ. The function $R_k \colon X \times Y \to \mathbb{K}$, $R_k(x, y) := \kappa(x, y) - \tilde{\kappa}(x, y)$, is called error term and it follows

$$\kappa(x, y) = \tilde{\kappa}(x, y) + R_k(x, y).$$

In this thesis, we will use polynomial interpolation to obtain good low rank approximations.

Chapter 3

Laplace transform and generalizations

In this chapter, we treat the computation of sums (1.1) in the one-dimensional case for so-called asymptotically smooth kernels. One example of such a kernel is the real valued exponential kernel. In this case, the sums (1.1) lead to a discrete Laplace transform. We study the fast Laplace transform [46] and improve slightly the error estimate of this approach. Moreover, we generalize this method to the computation of sums (1.1) for arbitrary asymptotically smooth kernels $\kappa\colon Y \times \Omega \to \mathbb{R}$, $Y, \Omega \subset [0, \infty)$.

To be more precise, let $M_1, M_2 \in \mathbb{N}$, evaluation or spatial nodes $y_1 > \ldots > y_{M_1} > 0$, frequency nodes $\xi_1 > \ldots > \xi_{M_2} > 0$, and coefficients $\hat{f}_k \in \mathbb{C}$ for $k = 1, \ldots, M_2$ be given. We consider the computation of sums

$$f_j := f(y_j) = \sum_{k=1}^{M_2} \hat{f}_k \kappa(y_j, \xi_k), \quad j = 1, \ldots, M_1. \tag{3.1}$$

Allowing only decreasing sequences y_j and ξ_k is just to simplify the notation. The computation of the sum is equivalent to compute the matrix vector product

$$\mathbf{f} = \mathbf{K}\hat{\mathbf{f}}, \quad \mathbf{K} := \big(\kappa(y_j, \xi_k)\big)_{j=1,k=1}^{M_1, M_2}, \tag{3.2}$$

where $\mathbf{f} = (f_j)_{j=1,\ldots,M_1}$ and $\hat{\mathbf{f}} = (\hat{f}_k)_{k=1,\ldots,M_2}$, respectively.

3.1 Asymptotically smooth kernels

At first, we introduce asymptotically smooth kernels similar to [29, Definition 4.2.5.]. Afterwards, we consider a low rank approximation by interpolation in Chebyshev nodes in both variables of such kernels.

Definition 3.1. *Let an infinitely often differentiable function* $\kappa\colon (0, \infty) \times (0, \infty) \to \mathbb{R}$ *be given. The kernel* κ *is said to be asymptotically smooth if there exist constants* $C, \mu, s \geq 0$, $\nu \in \mathbb{R}$ *such that for all* $q \in \mathbb{N}$ *the conditions*

$$\left| y^q \partial_y^q \kappa(y, \xi) \right| \leq Cq! \mu^q q^\nu (y\xi)^{-s} \quad \text{and} \quad \left| \xi^q \partial_\xi^q \kappa(y, \xi) \right| \leq Cq! \mu^q q^\nu (y\xi)^{-s}$$

are fulfilled for all $y, \xi \in (0, \infty)$. Moreover, we call two intervals $A, B \subset [0, \infty)$ admissible if

$$\operatorname{diam}(A) \leq \operatorname{dist}(A, 0) \quad and \quad \operatorname{diam}(B) \leq \operatorname{dist}(B, 0). \tag{3.3}$$

We give two examples of asymptotically smooth kernels, one for $s = 0$ and one for $s = 1/2$. Note that the parameter s characterizes the singularity of the kernel for $y\xi = 0$.

Lemma 3.2. *The kernel $\kappa : [0, \infty) \times [0, \infty) \to \mathbb{R}$,*

$$\kappa(y, \xi) := e^{-y\xi},$$

is asymptotically smooth with

$$\left| y^q \partial_y^q \kappa(y, \xi) \right| \leq \frac{q!}{\sqrt{2\pi q}} \quad and \quad \left| \xi^q \partial_\xi^q \kappa(y, \xi) \right| \leq \frac{q!}{\sqrt{2\pi q}}.$$

Proof. For $y > 0$, we have the necessary condition for a local maximum of $\left| \partial_y^q \kappa(y, \xi) \right|$, i.e.,

$$\left| \partial_\xi \partial_y^q \kappa(y, \xi) \right| = \xi^{q-1} e^{-\xi y} |q - \xi y| = 0$$

if and only if $\xi = q/y$. Using Stirling's approximation, we conclude the globally valid bound

$$\left| \partial_y^q \kappa(y, \xi) \right| = \xi^q e^{-\xi y} \leq \frac{q^q}{e^q y^q} \leq \frac{1}{\sqrt{2\pi q}} q! y^{-q}. \tag{3.4}$$

The proof is analogously for $\left| \partial_{\xi^q} \kappa(y, \xi) \right|$ and the assertion follows. $\qquad \square$

Lemma 3.3. *Let the modified Bessel function of the second kind $K_\eta \colon \mathbb{R}_+ \to \mathbb{R}$,*

$$K_\eta(x) := \int_0^\infty e^{-x \cosh(t)} \cosh(\eta t) \mathrm{d}t,$$

for $\eta \in \mathbb{R}$ be given. In particular, we have $K_{1/2}(x) = \sqrt{\frac{\pi}{2x}} e^{-x}$. Then, the kernel $\kappa : (0, \infty) \times (0, \infty) \to \mathbb{R}$,

$$\kappa(y, \xi) := K_{1/2}(y\xi)$$

is asymptotically smooth with

$$\left| y^q \frac{\partial^q}{\partial y^q} \kappa(y, \xi) \right| \leq \sqrt{\frac{\pi}{2y\xi}} q! \quad and \quad \left| \xi^q \frac{\partial^q}{\partial \xi^q} \kappa(y, \xi) \right| \leq \sqrt{\frac{\pi}{2y\xi}} q!.$$

Proof. The modified Bessel function for $\eta = \frac{1}{2}$ is stated in [56, 3.7.1, Formula (13)]. We prove per induction over $q \in \mathbb{N}_0$, that the q-th partial derivative of κ is given by

$$\frac{\partial^q}{\partial y^q} \kappa(y, \xi) = \sqrt{\frac{\pi}{2}} (-\xi)^q e^{-y\xi} \sum_{k=0}^q \binom{q}{k} \frac{\prod_{j=0}^{k-1}(2j+1)}{2^k} (y\xi)^{-(2k+1)/2}.$$

For $q = 0$ we have $\frac{\partial^q}{\partial y^q}\kappa(y,\xi) = \sqrt{\frac{\pi}{2y\xi}}e^{-y\xi}$. It follows

$$\partial_y^{q+1}\kappa(y,\xi) = \partial_y\partial_y^q\kappa(y,\xi))$$

$$= \sqrt{\frac{\pi}{2}}(-\xi)^{q+1}e^{-y\xi}\sum_{k=0}^{q}\binom{q}{k}\frac{\prod_{j=0}^{k-1}(2j+1)}{2^k}(y\xi)^{-(2k+1)/2}$$

$$+ \sqrt{\frac{\pi}{2}}(-\xi)^{q}e^{-y\xi}\sum_{k=0}^{q}\binom{q}{k}\frac{\prod_{j=0}^{k-1}(2j+1)}{2^k}\frac{-(2k+1)}{2}\xi(y\xi)^{-(2(k+1)+1)/2}$$

$$= \sqrt{\frac{\pi}{2}}(-\xi)^{q+1}e^{-y\xi}\left(\sum_{k=0}^{q}\binom{q}{k}\frac{\prod_{j=0}^{k-1}(2j+1))}{2^k}(y\xi)^{-(2k+1)/2}\right.$$

$$+ \left.\sum_{k=0}^{q}\binom{q}{k}\frac{\prod_{j=0}^{k}(2j+1)}{2^{k+1}}(y\xi)^{-(2(k+1)+1)/2}\right).$$

Shifting the index in the last sum $k \to k+1$ and applying $\binom{q+1}{k} = \binom{q}{k} + \binom{q}{k-1}$ implies the assertion. Furthermore, using $\prod_{j=0}^{k-1}(2j+1) \le \prod_{j=1}^{k}2j = 2^k k!$ and shifting $l = q - k$ yields

$$\left|\frac{\partial^q}{\partial y^q}\kappa(y,\xi)\right| \le \sqrt{\frac{\pi}{2y\xi}}\xi^q e^{-y\xi}\sum_{k=0}^{q}\binom{q}{k}k!(y\xi)^{-k}$$

$$= \sqrt{\frac{\pi}{2y\xi}}\xi^q e^{-y\xi}q!\sum_{k=0}^{q}\frac{1}{(q-k)!}(y\xi)^{-k}.$$

$$\le \sqrt{\frac{\pi}{2y\xi}}\xi^q e^{-y\xi}q!(y\xi)^{-q}\sum_{l=0}^{\infty}\frac{1}{l!}(y\xi)^{l}$$

$$= \sqrt{\frac{\pi}{2y\xi}}\xi^q q!(y\xi)^{-q}.$$

The multiplication with y^q implies the assertion for y. Since κ is symmetric, i.e. $\kappa(y,\xi) = \kappa(\xi,y)$, the proof can be done analogously for ξ. □

3.1.1 Low rank approximation via interpolation

Interpolation in Chebyshev nodes in both variables leads to a good low rank approximation of asymptotically smooth kernels under some admissibility condition. At first, we introduce the interpolation in a tensor Chebyshev grid.

Definition 3.4. *Let $A, B \subset (0,\infty)$, $\kappa\colon A \times B \to \mathbb{R}$, and $q \in \mathbb{N}$. We denote by $\mathcal{I}$ the identity, then we define the interpolations in one variable via*

$$(\mathcal{I}_q^A \otimes \mathcal{I})\kappa(y,\xi) := \sum_{s=0}^{q-1} L_s^A(y)\kappa(y_s^A,\xi),$$

$$(\mathcal{I} \otimes \mathcal{I}_q^B)\kappa(y,\xi) := \sum_{r=0}^{q-1} L_r^B(\xi)\kappa(y,\xi_r^B),$$

and in two variables via

$$\mathcal{I}_q^{A\times B}\kappa(y,\xi) := (\mathcal{I}_q^A \otimes \mathcal{I}_q^B)\kappa(y,\xi) := \sum_{s=0}^{q-1}\sum_{r=0}^{q-1} L_s^A(y)\kappa(y_s^A,\xi_r^B)L_r^B(\xi).$$

The interpolation obeys the following error estimate.

Theorem 3.5. *Let $q \in \mathbb{N}$, $q \geq 2$, $A, B \subset (0,\infty)$ be admissible, and $\kappa\colon A \times B \to \mathbb{R}$ be asymptotically smooth with constants $C, \mu, s \geq 0$ and $\nu \in \mathbb{R}$, then we have*

$$\left\|\kappa - \mathcal{I}_q^{A\times B}\kappa\right\|_{C(A\times B)} \leq \frac{C\mu^q q^\nu}{2^{2q-1}}\left(2 + \frac{2}{\pi}\log q\right)(\operatorname{dist}(A,0)\operatorname{dist}(B,0))^{-s}.$$

Proof. For fixed $\xi \in B$ and $g\colon A \to \mathbb{R}$, $g(y) := \kappa(y,\xi)$, we apply the error formula of Theorem 2.13 and obtain

$$\|g - \mathcal{I}_q^A g\|_{C(A)} \leq \frac{\operatorname{diam}(A)^q}{2^{2q-1}q!}\|g^{(q)}\|_{C(A)}.$$

The asymptotical smoothness and the admissibility implies

$$\left|g^{(q)}(y)\right| \leq Cq!\mu^q q^\nu y^{-q}(y\xi)^{-s} \leq Cq!\mu^q q^\nu(\operatorname{dist}(A,0))^{-q}(y\xi)^{-s}$$

and in conclusion

$$\|g - \mathcal{I}_q^A g\|_{C(A)} \leq C\mu^q q^\nu 2^{1-2q}\sup_{y\in A,\xi\in B}|(y\xi)^{-s}|.$$

The same estimate holds true with respect to $\xi \in B$. In particular, Definition 3.4 yields

$$(\mathcal{I}_q^A \otimes \mathcal{I})(\mathcal{I} \otimes \mathcal{I}_q^B)\kappa(y,\xi) = (\mathcal{I}_q^A \otimes \mathcal{I})\sum_{r=0}^{q-1}\kappa(y,\xi_r^B)L_r^B(\xi) = \mathcal{I}_q^{A\times B}\kappa(y,\xi).$$

From this and together with the Lebesgue constant in Corollary 2.15, we conclude

$$\begin{aligned}
\left\|\kappa - \mathcal{I}_q^{A\times B}\kappa\right\|_{C(A\times B)} &\leq \left\|\kappa - \left(\mathcal{I}_q^A \otimes \mathcal{I}\right)\kappa\right\|_{C(A\times B)} \\
&\quad + \left\|\mathcal{I}_q^A \otimes \mathcal{I}\right\|\left\|\kappa - \left(\mathcal{I} \otimes \mathcal{I}_q^B\right)\kappa\right\|_{C(A\times B)} \\
&\leq \frac{C\mu^q q^\nu}{2^{2q-1}}\left(2 + \frac{2}{\pi}\log q\right)\sup_{y\in A,\xi\in B}|(y\xi)^{-s}|.
\end{aligned}$$

The conditions $y\xi \geq \operatorname{dist}(A,0)\operatorname{dist}(B,0)$ and $s \geq 0$ imply the assertion. $\qquad\square$

3.2 Laplace transform

For the rest of this section, we consider the exponential kernel $\kappa(y, \xi) = e^{-\xi y}$. We call the sums (3.1) discrete Laplace transform. The following results are a simplification and minor improvement of [46].

Corollary 3.6 ([46, Equation (36)]). *Let $q \in \mathbb{N}$, $q \geq 2$, $A, B \subset [0, \infty)$ be admissible, then*

$$\left\| \kappa - \mathcal{I}_q^{A \times B} \kappa \right\|_{C(A \times B)} \leq 2^{1-2q}.$$

Proof. Due to Lemma 3.2, the exponential kernel is asymptotically smooth with constants $C = 1/\sqrt{2\pi}$, $\mu = 1$, $\nu = -1/2$, and $s = 0$. Let $f \colon (0, \infty) \to \mathbb{R}$,

$$f(q) := 2 + \frac{2}{\pi} \log q - \sqrt{2\pi q},$$

be given. Since f is continuous in $(0, \infty)$, $\lim_{q \to 0^+} f(q) = -\infty$ and has only one extremal point in $q = \frac{8}{\pi^3}$ which is a maximum, it follows $f(q) \leq f(\frac{8}{\pi^3}) \leq 0$. This implies $(2 + \frac{2}{\pi} \log q)/\sqrt{2\pi q} \leq 1$ and the assertion follows from Theorem 3.5. $\qquad\square$

Definition 3.7. *For given $\varepsilon, y_1, \xi_1 > 0$, we define the geometrical decomposition of $Y := [0, y_1]$ and $\Omega := [0, \xi_1]$ by*

$$M := \left\lceil \log_2 \frac{y_1 \xi_1}{\varepsilon} \right\rceil + 1, \qquad\qquad m := 1, \ldots, M - 1,$$

$$Y_M := \left[0, \frac{y_1}{2^{M-1}} \right], \qquad\qquad Y_m := \left(\frac{y_1}{2^m}, \frac{y_1}{2^{m-1}} \right],$$

$$\Omega_M := \left[0, \frac{\xi_1}{2^{M-1}} \right], \qquad\qquad \Omega_m := \left(\frac{\xi_1}{2^m}, \frac{\xi_1}{2^{m-1}} \right].$$

For ease of notation in Algorithm 1, we moreover define

$$\mathbf{L}^{\Omega_\ell} := \left(L_r^{\Omega_\ell}(\xi_j) \right)_{\xi_j \in \Omega_\ell, r=0, \ldots, q-1}, \qquad \mathbf{L}^{Y_m} := \left(L_s^{Y_m}(y_i) \right)_{y_i \in Y_m, s=0, \ldots, q-1},$$

$$\hat{\mathbf{f}}^{\Omega_\ell} := \left(\hat{f}_j \right)_{\xi_j \in \Omega_\ell}, \qquad \mathbf{K}^{Y_m, \Omega_\ell} := \left(\kappa(y_s^{Y_m}, \xi_r^{\Omega_\ell}) \right)_{s, r=0}^{q-1, q-1}$$

for $m, \ell = 1, \ldots, M - 1$.

Figure 3.1: Geometrical decomposition of the domain $Y = [0, y_1]$ for $M = 4$.

Lemma 3.8 ([46, Sect. 4]). *Let $\varepsilon, y_1, \xi_1 > 0$ be given, use the notation of Definition 3.7 and Corollary 3.6 and set $q := \lceil \frac{1}{2} + \log_4 1/\varepsilon \rceil$,*

$$\ell_m := \max(1, \lfloor \log_2(y_1 \xi_1) - m - \log_2(\log 1/\varepsilon) \rfloor + 1),$$

$$L_m := M - m$$

for all $m = 1, \ldots, M - 1$, then

(i) $y \in Y_M$ and $\xi \in \Omega$ *(and analogously $y \in Y$ and $\xi \in \Omega_M$) implies* $1 - e^{-y\xi} \leq \varepsilon$,

(ii) $y \in Y_m$, $m = 1, \ldots, M-1$, and $\xi \in \Omega_\ell$, $\ell < \ell_m$, *implies* $e^{-y\xi} \leq \varepsilon$,

(iii) $y \in Y_m$, $m = 1, \ldots, M-1$, and $\xi \in \Omega_\ell$, $\ell_m \leq \ell \leq L_m$, *implies*

$$\left| e^{-y\xi} - \mathcal{I}_q^{Y_m \times \Omega_\ell} \kappa(y, \xi) \right| \leq \varepsilon,$$

(iv) $y \in Y_m$, $m = 1, \ldots, M-1$, and $\xi \in \Omega_\ell$, $\ell > L_m$, *implies* $1 - e^{-y\xi} \leq \varepsilon$.

Proof. The individual estimates can be proven as follows. At first, let $\xi \in [0, \xi_1]$ and $y \in Y_M = [0, y_1/2^{M-1}]$. Using $M \geq \log_2 \frac{y_1 \xi_1}{\varepsilon} + 1$, we obtain $0 \leq \xi y \leq y_1 \xi_1 / 2^{M-1} \leq \varepsilon$ and finally case i) since

$$1 \geq e^{-y\xi} \geq e^{-\varepsilon} = \sum_{k=0}^{\infty} \frac{(-\varepsilon)^k}{k!} \geq 1 - \varepsilon.$$

Now let $y \in Y_m$, $\xi \in \Omega_\ell$. The condition $\ell \leq \ell_m - 1 = \lfloor \log_2(y_1 \xi_1) - m - \log_2(\log 1/\varepsilon) \rfloor$ implies

$$e^{-\frac{y_1 \xi_1}{2^m 2^\ell}} \leq \varepsilon$$

and due to $y \geq \frac{y_1}{2^m}$, $\xi \geq \frac{\xi_1}{2^\ell}$ assertion (ii). The third result follows from Corollary 3.6 since the intervals Y_m, Ω_ℓ, $\ell, m = 1, \ldots, M-1$, are admissible.

Finally, we have $y \in Y_m$, $\xi \in \Omega_\ell$, $\ell - 1 \geq L_m = M - m = \lceil \log_2(\xi_1 y_1) + \log_2(1/\varepsilon) \rceil - (m-1)$ and thus

$$1 - e^{-y\xi} \leq 1 - e^{-\frac{y_1}{2^{m-1}} \frac{\xi_1}{2^{\ell-1}}} \leq 1 - e^{-\varepsilon} \leq \varepsilon.$$

$\square$

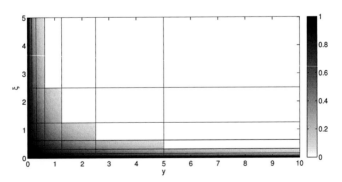

Figure 3.2: Contour plot of the Kernel function and geometrical decomposition, the cases in Lemma 3.8 (i), (ii), and (iv) are shown in black and white, respectively.

Theorem 3.9 ([46, Sect. 7]). *Let* $M_1, M_2 \in \mathbb{N}$, $\varepsilon > 0$, $y_1 > y_2 > \ldots > y_{M_1} > 0$, $\xi_1 > \xi_2 > \ldots > \xi_{M_2} > 0$, $\hat{\mathbf{f}} \in \mathbb{C}^{M_2}$, *and* $\mathbf{f} = \mathbf{K}\hat{\mathbf{f}}$ *be given. Set* $M := \lceil \log_2 \frac{y_1 \xi_1}{\varepsilon} \rceil + 1$, $q := \lceil \frac{1}{2} + \log_4 1/\varepsilon \rceil$, *and* $\widetilde{\mathbf{f}} = (\widetilde{f}(y_i))_{i=1,\ldots,M_1}$ *for the function* $\widetilde{f} : Y \to \mathbb{C}$,

$$\widetilde{f}(y) = \begin{cases} \sum_{j=1}^{M_2} \hat{f}_j & y \in Y_M, \\ \sum_{\ell=\ell_m}^{L_m} \sum_{\xi_j \in \Omega_\ell} \hat{f}_j \mathcal{I}^{Y_m \times \Omega_\ell} \kappa(y, \xi_j) + \sum_{\ell > L_m} \sum_{\xi_j \in \Omega_\ell} \hat{f}_j & y \in Y_m, \ 1 \le m < M. \end{cases} \tag{3.5}$$

Then the error estimate

$$\|\mathbf{f} - \widetilde{\mathbf{f}}\|_\infty \le \varepsilon \|\hat{\mathbf{f}}\|_1$$

holds true and Algorithm 1 computes this approximation in

$$\mathcal{O}\left((M_1 + M_2) \log \frac{1}{\varepsilon} + \left(\log^3 \frac{1}{\varepsilon} \right) \log \frac{y_1 \xi_1}{\varepsilon} \right) \tag{3.6}$$

floating point operations.

Proof. We start with the error estimate. For $y \in Y_M$, Lemma 3.8 (i) implies

$$\left| f(y) - \widetilde{f}(y) \right| \le \sum_{k=1}^{M_2} |\hat{f}_k| |e^{-\xi_k y} - 1| \le \varepsilon \sum_{k=1}^{N} |\hat{f}_k|.$$

Now let $m = 1, \ldots, M - 1$, $y \in Y_m$, and partition the function f in three parts

$$f(y) = \left(\sum_{\ell < \ell_m} + \sum_{\ell=\ell_m}^{L_m} + \sum_{\ell > L_m} \right) \sum_{\xi_j \in \Omega_\ell} \hat{f}_j \kappa(y, \xi_j).$$

The desired result follows by the application of Lemma 3.8 (ii) – (iv) and the approximation of the kernel κ by zero, by interpolation, or by one, respectively.

Regarding the computational costs, we first note that the last sum in Equation (3.5), denoted by g_m in Algorithm 1, is computed for all $m = 1, \ldots, M$ by means of a cumulative summation in $\mathcal{O}(M_2)$ operations. In matrix notation, the spatial partitions in Definition 3.7 yield a partition of the matrix in Equation (3.2) in admissible blocks. Considering only the case when the kernel function is approximated by interpolation, Algorithm 1 factors out row- and column bases by

$$\tilde{\mathbf{K}} = \left(\mathbf{L}^{Y_m} \mathbf{K}^{Y_m, \Omega_\ell} \left(\mathbf{L}^{\Omega_\ell} \right)^\top \right)_{m=1, \ell=\ell_m}^{M-1, L_m} \tag{3.7}$$

$$= \begin{pmatrix} \mathbf{L}^{Y_1} & 0 & 0 \\ 0 & \ddots & 0 \\ 0 & 0 & \mathbf{L}^{Y_{M-1}} \end{pmatrix} \begin{pmatrix} \ddots & \vdots & \cdot^{\cdot^\cdot} \\ \cdots & \mathbf{K}^{Y_m, \Omega_\ell} & \cdots \\ \cdot_{\cdot_\cdot} & \vdots & \ddots \end{pmatrix}_{m=1, \ell=\ell_m}^{M-1, L_m} \begin{pmatrix} \left(\mathbf{L}^{\Omega_{\ell_m}} \right)^\top & 0 & 0 \\ 0 & \ddots & 0 \\ 0 & 0 & \left(\mathbf{L}^{\Omega_{L_m}} \right)^\top \end{pmatrix}.$$

Clearly, applying the rightmost block diagonal matrix takes at most $\sum_{\ell=\ell_m}^{L_m} q|\Omega_\ell| \le qM_2$ operations. The second matrix has at most $L_m - \ell_m \le 2\log(1/\varepsilon)$ blocks in its m-th block row, in total its application takes $\mathcal{O}(Mq^2 \log \frac{1}{\varepsilon})$ operations and the multiplication with the left block diagonal matrix takes at most qM_1 operations. Finally, note that we neglect the pre-computation of the necessary matrices $\mathbf{L}^{\Omega_\ell}$, $\mathbf{L}^{Y_m}$, and $\mathbf{K}^{Y_m,\Omega_\ell}$. This sums up to a complexity of $\mathcal{O}((M_1 + M_2) \log \frac{1}{\varepsilon} + Mq^2 \log \frac{1}{\varepsilon})$. Applying $M \in \mathcal{O}(\log \frac{y_1 \xi_1}{\varepsilon})$ and $q \in \mathcal{O}(\log \frac{1}{\varepsilon})$ completes the proof. $\qquad\qquad\qquad\qquad\qquad\qquad\qquad\square$

Algorithm 1 Laplace transform

Input:

$\quad \varepsilon \in (0,1)$ $\qquad\qquad\qquad\qquad\qquad\qquad\qquad\qquad\qquad\qquad\qquad$ ▷ target accuracy

$\quad N \in \mathbb{N}$ $\qquad\qquad\qquad\qquad\qquad\qquad\qquad\qquad\qquad\qquad$ ▷ number of sampling nodes

$\quad \xi_1 > \xi_2 \cdots > \xi_{M_2} > 0$ $\qquad\qquad\qquad\qquad\qquad$ ▷ nodes in frequency domain

$\quad y_1 > y_2 > \cdots > y_{M_1} > 0$ $\qquad\qquad\qquad\qquad$ ▷ nodes in spatial domain

$\quad \hat{\mathbf{f}} = (\hat{f}_j)_{j=1}^{M_2} \in \mathbb{C}^{M_2}$ $\qquad\qquad\qquad\qquad\qquad$ ▷ Fourier coefficients

Output:

$\quad \widetilde{\mathbf{f}} \in \mathbb{C}^N, \ \widetilde{\mathbf{f}} \approx \mathbf{K}\hat{\mathbf{f}}$ $\qquad\qquad\qquad\qquad\qquad$ ▷ samples in spatial domain

---

$\quad M = \left\lceil \log_2 \frac{y_1 \xi_1}{\varepsilon} \right\rceil + 1$ $\qquad\qquad\qquad\qquad$ ▷ number of decompositions

$\quad q = \left\lceil \frac{1}{2} + \log_4 1/\varepsilon \right\rceil$ $\qquad\qquad\qquad\qquad\qquad$ ▷ approximation rank

$\quad g_M = \sum_{\xi_j \in \Omega_M} \hat{f}_j$

$\quad$ **for** $m = M - 1, \ldots, 1$ **do**

$\qquad g_m = g_{m+1} + \sum_{\xi_j \in \Omega_m} \hat{f}_j$

$\quad$ **end for**

$\quad \widetilde{\mathbf{f}}^{Y_M} = g_1$

$\quad$ **for** $\ell = 1, \ldots, M - 1$ **do**

$\qquad \mathbf{v}^{\Omega_\ell} = \left(\mathbf{L}^{\Omega_\ell}\right)^\top \hat{\mathbf{f}}^{\Omega_\ell}$

$\quad$ **end for**

$\quad$ **for** $m = 1, \ldots, M - 1$ **do**

$\qquad \mathbf{h}^{Y_m} = \sum_{\ell=\ell_m}^{L_m} \mathbf{K}^{Y_m,\Omega_\ell} \mathbf{v}^{\Omega_\ell}$

$\qquad \widetilde{\mathbf{f}}^{Y_m} = \mathbf{L}^{Y_m} \mathbf{h}^{Y_m} + g_{L_m+1} \mathbf{1}$

$\quad$ **end for**

Remark 3.10. *The approximation in Theorem 3.9 yields an entrywise error*

$$\|\mathbf{K} - \tilde{\mathbf{K}}\|_{1\to\infty} = \max_{j,l} |\mathbf{K}_{j,l} - \tilde{\mathbf{K}}_{j,l}| \le \varepsilon.$$

In particular, this implies

$$\|\mathbf{K} - \tilde{\mathbf{K}}\|_2 \le \|\mathbf{K} - \tilde{\mathbf{K}}\|_F \le N\|\mathbf{K} - \tilde{\mathbf{K}}\|_{1\to\infty} \le N\varepsilon.$$

and introduces at most a $\log N$*-factor in the complexity estimate if we wish to achieve accuracy* ε *with respect to the spectral norm. We note in passing, that the adaptive cross approximation [4] of the matrix block* $\mathbf{K}^{Y_m,\Omega_\ell}$ *takes the form*

$$\bar{\mathbf{K}} = \mathbf{K}^{Y_m} \tilde{\mathbf{K}}_q^{-1} \mathbf{K}^{\Omega_\ell},$$

where all matrices on the right hand side evaluate the kernel κ at certain nodes. In case of total pivoting, this leads to an error $\|\mathbf{K} - \bar{\mathbf{K}}\|_{1\to\infty} \leq (q+1)\sigma_{q+1}$, where σ_q denotes the q-th singular value of $\mathbf{K}$, cf. [22].

An application of Algorithm 1 is the evaluation of certain polynomials.

Remark 3.11. *(i) Algorithm 1 can evaluate polynomials $p \in \Pi_{M_2}(\mathbb{R}, \mathbb{C})$ of the form*

$$p(x) = \sum_{k=1}^{M_2} \hat{f}_k x^k$$

in $0 < x_1 < \cdots < x_{M_1} < 1$, where $M_1, M_2 \in \mathbb{N}$. We set $y_j := -\log(x_j)$, which reduces the evaluation of p to an evaluation of exponentials in the decreasing sequence y_j. In this case, the complexity (3.6) can be written as

$$\mathcal{O}\left((M_1 + M_2)\log\frac{1}{\varepsilon} + \log^3\frac{1}{\varepsilon}\log\left(\frac{|\log x_1|M_2}{\varepsilon}\right)\right).$$

See also [46, Remark 3.1].

(ii) Let $K, M_1, M_2 \in \mathbb{N}$ be given. Moreover, let

$$0 < x_1 < \cdots < x_{M_1} < 1 \quad and \quad K \geq \xi_1 > \cdots > \xi_{M_2} > 0$$

with $\xi_k \in \mathbb{N}$ be given. Another approach is the evaluation of sparse polynomials $p \in \Pi_K(\mathbb{R}, \mathbb{C})$ of the form

$$p(x) = \sum_{k=1}^{M_2} \hat{f}_k x^{\xi_k}.$$

Setting again $y_j := -\log(x_j)$, we can compute the values $p(x_j)$ for $j = 1, \ldots, M_1$ by Algorithm 1 in $\mathcal{O}((M_1 + M_2)\log\frac{1}{\varepsilon} + \log^3\frac{1}{\varepsilon}\log(\frac{|\log x_1|K}{\varepsilon}))$ floating point operations.

Remark 3.12. *The truncated Laplace transform is given by*

$$\mathcal{L}_0 \colon L^2[1,2] \to L^2[1,2], \quad f \mapsto g(\xi) = \int_1^2 f(x)e^{-\xi x}dx$$

and [39, Equation (3.105)] conjectures a singular value decay

$$\sigma_q \approx C\sqrt{2\pi}e^{-\sqrt{10.1189q(1+q)+6.12572}},$$

which serves as an asymptotic lower bound on the best approximation of the real exponential kernel under the admissibility condition (3.3). A comparison between the singular value decay and the local error is illustrated in Figure 3.3.

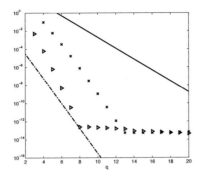

Figure 3.3: The ℓ^2-approximation error (daggers) of the polynomial interpolation, see
Definition 3.4, together with its normalized theoretical upper bound (solid
line) given in Corollary 3.6, and the singular values (triangles), representing the best approximation in this case, together with the lower bound of
Remark 3.12 (dashed line).

3.3 Special function transforms

The Laplace transform can be extended to asymptotically smooth kernels. The generalization of Algorithm 1 and Theorem 3.9 to other asymptotically smooth kernels reads
as follows. We suppose the kernel κ is asymptotically smooth with constants $C > 0$,
$\nu \in \mathbb{R}$, $s \geq 0$ and $\mu \in (0,4)$, which implies a local approximation

$$\left\| \kappa - \mathcal{I}_q^{A \times B} \kappa \right\|_{C(A \times B)} \leq \tilde{C} c^q (\operatorname{dist}(A,0) \operatorname{dist}(B,0))^{-s},$$

with some constant $\tilde{C} \geq 0$ and $c \in (0,1)$. A bound for the interpolation error away
from the singularity $y\xi = 0$ is given in the next corollary. We increase the number of
boxes M by a $\log N$ factor in the geometrical decompositions to obtain under some
further assumptions on the nodes later a constant number of nodes in the near fields
Y_M and Ω_M.

Corollary 3.13. *Let* $M := \left\lceil \log_2 \frac{y_1 \xi_1 N}{\varepsilon} \right\rceil + 1$ *and* $q \geq \left\lceil \frac{\log(\varepsilon^{2s+1}/((2y_1\xi_1)^s N^{2s}\tilde{C}))}{\log c} \right\rceil$ *be given.*
For $A \neq Y_M$ *and* $B \neq \Omega_M$, *it follows*

$$\left\| \kappa - \mathcal{I}_q^{A \times B} \kappa \right\|_{C(A \times B)} \leq \varepsilon.$$

Proof. We have

$$\left\| \kappa - \mathcal{I}_q^{A \times B} \kappa \right\|_{C(A \times B)} \leq \tilde{C} c^q (\operatorname{dist}(A,0) \operatorname{dist}(B,0))^{-s}.$$

Since $M - 1 \leq \log_2 \frac{y_1 \xi_1 N}{\varepsilon} + \frac{1}{2}$, it follows

$$\operatorname{dist}(A,0) \operatorname{dist}(B,0) \geq \frac{y_1}{2^{M-1}} \frac{\xi_1}{2^{M-1}} \geq \frac{\varepsilon^2}{2y_1 \xi_1 N^2}. \tag{3.8}$$

This implies

$$\left\| \kappa - \mathcal{I}_q^{A \times B} \kappa \right\|_{C(A \times B)} \leq \frac{\tilde{C} c^q}{(\text{dist}(A,0) \, \text{dist}(B,0))^s}$$

$$\leq \tilde{C} (2 y_1 \xi_1)^s \left(\frac{N}{\varepsilon} \right)^{2s} c^q$$

and with $q \geq \left| \frac{\log(\varepsilon^{2s+1}/((2y_1 \xi_1)^s N^{2s} \tilde{C}))}{\log c} \right|$ the assertion. $\qquad \square$

Example 3.14. *Let κ be given by the modified Bessel function of the second kind, $\kappa(x,y) = \frac{1}{\sqrt{2\pi x y}} e^{-y\xi}$, see Lemma 3.3. In other words, we have $s = \frac{1}{2}$. At first, we determine the constants $\tilde{C}$ and c in the previous corollary. Combining the asymptotical smoothness, Lemma 3.3, and Theorem 3.5 yields*

$$\left\| \kappa - \mathcal{I}_q^{A \times B} \kappa \right\|_{C(A \times B)} \leq \sqrt{2\pi} \left(2 + \frac{2}{\pi} \log q \right) \frac{1}{4^q} (\text{dist}(A,0) \, \text{dist}(B,0))^{-1/2}.$$

From the proof of Corollary 3.6 it follows

$$\frac{\sqrt{2\pi}(2 + \frac{2}{\pi} \log q)}{4^q} \leq \frac{2\pi \sqrt{q}}{4^q} \leq \frac{2\pi}{3^q},$$

where the last estimate can be proven inductively. This leads to constants $\tilde{C} = 2\pi$ and $c = \frac{1}{3}$. The approximation rank has to fulfill $q \geq 2 \left| \log \frac{\varepsilon}{\sqrt[4]{8 N^2 \pi^2 y_1 \xi_1}} \right|$ to obtain a target accuracy ε.

Remark 3.15. *The generalization of Algorithm 1 and Theorem 3.9 to other asymptotically smooth kernels reads as follows. Let the assumptions of the previous Corollary be given. Moreover, let the nodes*

$$y_1 > y_2 > \ldots > y_N > 0 \quad and \quad \xi_1 > \xi_2 > \ldots > \xi_N > 0$$

be quasi-uniform and $y_1, \xi_1 > 0$ be fixed for all $N \in \mathbb{N}$. Since the domains Y and Ω are geometrically decomposed, there are $\mathcal{O}(\frac{N}{2^{M-1}})$ nodes in the near fields Y_M and Ω_M. Applying $M \in \mathcal{O}(\log \frac{N}{\varepsilon})$ leads to a constant number $(\mathcal{O}(\varepsilon))$ of nodes in the near fields. We apply direct computations in the near fields Y_M and Ω_M. Of course the approximations of the kernel by either zero or one as in Lemma 3.8 cannot be done in general and thus we let run $\ell = 1, \ldots, M - 1$ in Lemma 3.8 (iii). We obtain for $f(y) = \sum_{k=1}^{N} \hat{f}_k \kappa(y, \xi_k)$ an approximation

$$\widetilde{f}(y) = \begin{cases} f(y) & y \in Y_M, \\ \sum_{\ell=1}^{M-1} \sum_{\xi_j \in \Omega_\ell} \hat{f}_j \mathcal{I}^{Y_m \times \Omega_\ell} \kappa(y, \xi_j) + \sum_{\xi_k \in \Omega_M} \hat{f}_k \kappa(y, \xi_k) & y \in Y_m, \ 1 \leq m < M. \end{cases}$$

Algorithm 1 then takes $\mathcal{O}(N\varepsilon)$ floating point operations for the exact computations in the near fields and $\mathcal{O}(qN + M^2 q^2)$ operations for the approximation, see Equation (3.7). Applying $q \in \mathcal{O}(\log \frac{N}{\varepsilon})$ the total complexity is reduced to $\mathcal{O}(N \log \frac{N}{\varepsilon} + \log^4 \frac{N}{\varepsilon})$ floating point operations.

3.4 Numerical experiments

The implementation of Algorithm 1 is realized in MATLAB 2013a. We use one node of an Intel Xeon, 128GByte, 2.2GHz, Scientific Linux release 6.5 (Carbon) for all numerical experiments.
We draw random uniformly distributed coefficients $\hat{f}_k$, and frequencies $\xi_k \in [0, N]$ for $k = 1, \ldots, N$, and we draw random nodes $0 \leq y_N \leq y_{N-1} \leq \cdots \leq y_1 \leq (2q - 1)\log 2$, which ensures $y_j \in [0, \log 1/\varepsilon]$ and $M = \lceil \log_2 \frac{N \log 1/\varepsilon}{\varepsilon} \rceil + 1$.
We consider the relative error

$$\varepsilon_1 := \frac{\|\mathbf{f} - \widetilde{\mathbf{f}}\|_\infty}{\|\hat{\mathbf{f}}\|_1}, \tag{3.9}$$

where $\mathbf{f}, \widetilde{\mathbf{f}} \in \mathbb{C}^N$ denote the exact result and its approximation, respectively, see Theorem 3.9. Figure 3.4 shows the quantity ε_1 and the corresponding upper bound in dependence of the approximation rank $q = 1, \ldots, 20$ for a fixed bandwidth $N = 2^{14}$.
In a second series of experiments we compare the computational times, measured by the MATLAB functions `tic` and `toc`, of the naive evaluations and Algorithm 1 with respect to the increasing problem size N. The theoretical complexity in (3.6) reduces to $\mathcal{O}(N \log \frac{1}{\varepsilon} + \log^3 \frac{1}{\varepsilon} \log \frac{N \log 1/\varepsilon}{\epsilon})$ for our choice of sampling nodes. For large values of N, the linear term in N dominates the complexity. Figure 3.5 shows the timings for the naive matrix vector multiplication using entrywise and rowwise evaluations of the matrix, both shown as diamonds, and Algorithm 1 is shown as plus. The complexity of the approximate computation is linearly in the problem size as shown in Figure 3.5.

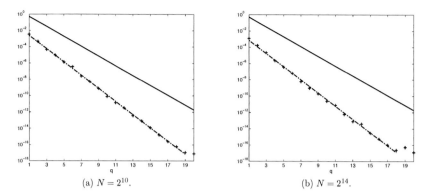

Figure 3.4: Approximation error ε_1 with respect to the local expansion degree q for a fixed bandwidth $N = 2^{10}$ (left) and $N = 2^{14}$ (right). The daggers represent the numerical errors, the solid line the theoretical estimate, cf. Corollary 3.6, and the dashed line a least square fit $\varepsilon_1 \approx C_0 C^{-q}$, $C > 4$.

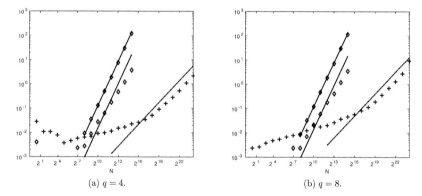

Figure 3.5: Behavior of the computational times of the naive computation (diamonds) and Algorithm 1 (plus) for fixed approximation rank $q = 4$ (left) and $q = 8$ (right) with respect to the bandwidth N. The solid lines represents a least square fit $t \approx C_1 N^2$ to the times t of the exact computations and the dashed lines a least square fit $t \approx C_2 N$ to the times t of the approximations.

Chapter 4

Butterfly sparse fast Fourier transform

Several applications in numerical analysis, signal processing, and imaging require the computation of the well known fast Fourier transform (FFT). The FFT is a divide and conquer algorithm which computes the coefficients of the trigonometric interpolation polynomial for equispaced interpolation nodes in $\mathcal{O}(N \log N)$ floating point operations. The inverse FFT allows for the evaluation of the interpolation polynomial in equispaced nodes. The evaluation of trigonometric polynomials in nonequispaced nodes led to the NFFT [12, 11, 5, 48, 15, 33]. Moreover, the NNFFT [15] allows for the evaluation of a non-harmonic function given by linear combinations of complex exponential functions at nonequispaced frequencies in nonequispaced spatial nodes. Let now N be large and a lot of Fourier coefficients be zero, neither the FFT nor its nonequispaced variants do take any advantage of this sparse setting. One idea to overcome this issue is the pruned FFT [19] which reduces the complexity for S nonzero Fourier coefficients to $\mathcal{O}(N \log S)$ floating point operations instead of $\mathcal{O}(NS)$ operations. A straightforward discretization to higher dimensions leads to a serious growth in computational costs.

In this chapter, we present a fast Fourier transform for certain sparse discretizations. To be more precise, let the bandwidth $N := 2^L$, $L \in \mathbb{N}$, and a dimension $d \in \mathbb{N}$ be given. Furthermore, we define the spatial and frequency domain $X := \Omega := [0, N]^d$, respectively. We call $\widetilde{X} := \{\boldsymbol{x}_j \in X : j = 1, \ldots, M_1\}$ sampling nodes in the spatial domain and $\widetilde{\Omega} := \{\boldsymbol{\xi}_k \in \Omega : k = 1, \ldots, M_2\}$ sampling nodes in the frequency domain, where $M_1, M_2 \in \mathbb{N}$. The task is the computation of the sums

$$f_j := f(\boldsymbol{x}_j) = \sum_{k=1}^{M_2} \hat{f}_k e^{2\pi i \boldsymbol{\xi}_k \boldsymbol{x}_j / N}, \quad j = 1, \ldots, M_1, \tag{4.1}$$

for a set of coefficients $\hat{f}_k \in \mathbb{C}$, $k = 1, \ldots, M_2$, which is equivalent the computation of the matrix vector product

$$\mathbf{f} = \mathbf{F}\hat{\mathbf{f}}, \quad \mathbf{F} := \left(e^{2\pi i \boldsymbol{x}_j \xi_k / N}\right)_{j=1, k=1}^{M_1, M_2}, \tag{4.2}$$

where $\mathbf{f} = (f_j)_{j=1, \ldots, M_1}$ and $\hat{\mathbf{f}} = (\hat{f}_k)_{k=1, \ldots, M_2}$, respectively. We use a so-called butterfly scheme to approximate nonharmonic exponential sums. This scheme relies on a low rank approximation of the kernel function under some admissibility condition. The butterfly scheme is a multilevel approximation scheme and can be traced back at least to [42] and has found a series of recent applications in [60, 8, 43, 54, 10]. Moreover, it is well known that certain blocks of the discrete Fourier transform are approximately of low rank [38, 57, 13]. Low rank approximation of high frequency exponential sums has led to the butterfly sparse fast Fourier transforms [2, 58].

We follow the ideas of [58] and introduce the so called butterfly sparse fast Fourier transform in the univariate case and afterwards for the multivariate case of the trigonometric sum (4.1). This chapter is based on the authors diploma thesis [41] which treats a preliminary version of the butterfly fast Fourier transform, and on our paper [35]. Here, we present a local error analysis in Theorem 4.8 which relies directly on a Taylor expansion in the complex plane, given in Lemma 4.7. A full error analysis is neither contained in [58] nor in [42]. Note that a full error analysis is published in [10] for a butterfly algorithm for the synthetic aperture radar imaging which leads to a generalized Fourier sum in the two-dimensional case. The here presented full error analysis, cf. Theorem 4.22, is more specific and given for all dimensions. Moreover, we present two different approaches for the computation of local interpolating functions. On the one hand, we consider the original approach [58], the monomial type representation, and on the other hand a new variant, the Lagrange type representation. For both variants, we came up with stability results, cf. Theorems 4.16 and 4.15 based on [41, Satz 43] and [41, Folgerung 58], respectively. In Section 4.4, we consider the time and space complexity for all variants.

4.1 The univariate case

First, we consider a local approximation of the kernel via interpolation in Chebyshev nodes under some admissibility condition. The main idea is to pre-factor out the oscillatory component, interpolate the remaining nonoscillatory part, and finally remodulate the outcome.

4.1.1 Low rank approximation via interpolation

We introduce function spaces of complex exponential functions. Furthermore, we define a trigonometric interpolation operator for such complex exponential functions.

Definition 4.1. *For boxes $A, B \subset \mathbb{R}$ we define the linear space of all finite expansions of exponential functions*

$$E_B(A) := \left\{ g \colon A \to \mathbb{C} : g(x) = \sum_{j=1}^{K} \hat{g}_j e^{2\pi i \xi_j x / N}, \ K \in \mathbb{N}, \ \hat{g}_j \in \mathbb{C}, \ \xi_j \in B \right\},$$

the equispaced nodes

$$\xi_j^B := c^B + \beta_j \operatorname{diam} B, \quad where \quad \beta_j := \frac{1}{2} - \frac{j}{p-1}, \ j = 0, \ldots, p-1, \tag{4.3}$$

and the subspace with $p \in \mathbb{N}$ equispaced frequencies

$$E_B^p(A) := \left\{ g : A \to \mathbb{C} : g(x) = \sum_{j=0}^{p-1} \hat{g}_j^{AB} \mathrm{e}^{2\pi \mathrm{i} \xi_j^B x/N}, \; \hat{g}_j^{AB} \in \mathbb{C} \right\}.$$

Moreover, we define for $p \in \mathbb{N}$, $p \geq 2$, the trigonometric interpolation operator

$$\mathcal{J}_p^{AB} : C(A) \to E_B^p(A), \quad g \mapsto \mathcal{J}_p^{AB} g = \sum_{s=0}^{p-1} \hat{g}_s^{AB} \mathrm{e}^{2\pi \mathrm{i} x \xi_s^B/N}, \tag{4.4}$$

such that in the Chebyshev nodes $\boldsymbol{x}^A \in \mathbb{R}^p$ the interpolation condition

$$\mathcal{J}_p^{AB} g(x_r^A) = g(x_r^A) \quad \text{for} \quad r = 0, \dots, p-1$$

holds true. Moreover, we call two intervals $A, B \subset \mathbb{R}$ admissible if

$$\operatorname{diam} A \operatorname{diam} B \leq N \tag{4.5}$$

holds true.

By the next definition, we introduce an interpolation on an arc in the complex plane and show later an essential connection between Definition 4.1 and 4.2.

Definition 4.2. *Let boxes $A, B \subset \mathbb{R}$ and $p \in \mathbb{N}$, $p \geq 2$, be given. Moreover, let the mapping $z \colon A \to \mathbb{C}$,*

$$z(x) := \mathrm{e}^{-2\pi \mathrm{i} x \frac{\operatorname{diam} B}{(p-1)N}}, \tag{4.6}$$

and the Chebyshev nodes $z_j^{AB} := z(x_j^A)$ mapped on the arc $\Gamma_p^{AB} := z(A)$ be given. Under the admissible condition in the sense of Inquality (4.5), the nodes are distinct and we define the Lagrange polynomials $\widetilde{L}_k^{AB} \colon \Gamma_p^{AB} \to \mathbb{C}$ as well as the interpolation operator $\mathcal{I}_p^{AB} \colon C(\Gamma_p^{AB}) \to \Pi_{p-1}(\mathbb{C})$,

$$\widetilde{L}_k^{AB}(z) := \prod_{\substack{j=0 \\ j \neq k}}^{p-1} \frac{z - z_j^{AB}}{z_k^{AB} - z_j^{AB}}, \tag{4.7}$$

$$\mathcal{I}_p^{AB} \widetilde{g} := \sum_{j=0}^{p-1} \widetilde{g}(z_j^{AB}) \widetilde{L}_j^{AB}. \tag{4.8}$$

Before we come up with an existence, uniqueness and boundedness theorem of the trigonometric interpolation, we need the following estimates.

Lemma 4.3. *For $p \in \mathbb{N}$, $p \geq 3$, and $x \in \mathbb{R}$, $|x| \leq \frac{2\pi}{p-1}$, we have*

$$\cos \frac{2\pi}{p-1} \leq \cos x, \tag{4.9}$$

$$1 - \frac{x^2}{2} \leq \cos x, \tag{4.10}$$

$$\left(1 - \cos \frac{2\pi}{p-1}\right) \frac{(p-1)^2}{2\pi^2} \leq \frac{2(1 - \cos x)}{x^2}, \tag{4.11}$$

$$\frac{4}{\pi^2} \leq \left(\frac{2(1 - \cos x)}{x^2}\right)^{\frac{\pi}{x}} \xrightarrow{x \to 0} 1. \tag{4.12}$$

Proof. The first estimate follows since the cosine is decreasing in $[0, \frac{2\pi}{p-1}] \subset [0, \pi]$ and even. Integrating $\cos x \leq 1$ twice yields the second claim. The relations $x \leq \tan x$ for $x \in [0, \frac{\pi}{2})$ and $\cos x \leq 0$, $\sin x \geq 0$, $x \geq 0$ for $x \in [\frac{\pi}{2}, \pi]$, yield $x \cos x - \sin x \leq 0$ and by integration

$$x \sin x - 2(1 - \cos x) \leq 0, \qquad x \in [0, \pi].$$

Dividing by $x^3/2$, $x \neq 0$, together with $\lim_{x \to 0} f'(x) = 0$ this yields $f'(x) \leq 0$ for the even function $f(x) = \frac{2(1 - \cos x)}{x^2}$ and thus $f(x) \geq f\left(\frac{2\pi}{p-1}\right)$, $|x| \leq \frac{2\pi}{p-1}$, which is the third assertion. Considering $x \in [-\pi, 0)$, we have $0 < f(x) \leq 1$, which implies $(f(x))^{\frac{\pi}{x}} \geq 1$ for $x \in [-\pi, 0)$ and with L'Hospitals rule

$$\frac{\pi}{x} \log f(x) \to 0, \text{ for } x \to 0.$$

Setting $g(x) = \frac{x}{\pi} \log \frac{4}{\pi^2}$ and $h(x) = \log f(x)$ yields $g(0) = h(0)$, $g(\pi) = h(\pi)$, and

$$h''(x) = -\frac{1}{1 - \cos x} + \frac{2}{x^2} \leq 0, \quad x \in [0, \pi].$$

Hence, the function h is concave and we obtain $h(x) \geq g(x)$. Since g is decreasing, this yields $\frac{\pi}{x} h(x) \geq \log \frac{4}{\pi^2}$ and finally the assertion $(f(x))^{\frac{\pi}{x}} \geq \frac{4}{\pi^2}$ for $x \in [-\pi, \pi]$. $\qquad \square$

Lemma 4.4. *Let $p \in \mathbb{N}$, $p \geq 3$, two boxes $A, B \subset \mathbb{R}$ be admissible, and the Lagrange functions $l_r^{AB} \colon A \to \mathbb{C}$,*

$$l_r^{AB}(x) := \widetilde{L}_r^{AB}(z(x)), \quad r = 0, \ldots, p-1, \tag{4.13}$$

be given. Then, the interpolation (4.4) is unique, has the representation

$$\mathcal{J}_p^{AB} g(x) = e^{2\pi i \left(c^B + \frac{\operatorname{diam} B}{2}\right) x/N} \sum_{r=0}^{p-1} g(x_r^A) e^{-2\pi i \left(c^B + \frac{\operatorname{diam} B}{2}\right) x_r^A/N} l_r^{AB}(x), \tag{4.14}$$

and its operator norm is bounded by

$$\|\mathcal{J}_p^{AB}\| := \sup_{g \in C(A) \setminus \{0\}} \frac{\|\mathcal{J}_p^{AB} g\|_{C(A)}}{\|g\|_{C(A)}} \leq C_p, \quad C_p := K_p \left(1 + \frac{2}{\pi} \log p\right), \tag{4.15}$$

where

$$K_p := \left(\frac{2\pi^2}{\left(1 - \cos \frac{2\pi}{p-1}\right)(p-1)^2}\right)^{\frac{p-1}{2}}, \quad 1 \leq K_p \leq \frac{\pi^2}{4}, \quad \lim_{p \to \infty} K_p = 1. \tag{4.16}$$

Proof. We set the function values $\widetilde{g}(z_r^{AB}) = \widetilde{g}(z(x_r^A)) := e^{-2\pi i \left(c^B + \frac{\operatorname{diam} B}{2}\right) x_r^A/N} g(x_r^A)$ for $r = 0, \ldots, p-1$. The polynomial, which interpolates the points

$$(z_r^{AB}, \widetilde{g}(z_r^{AB})) \in \mathbb{C} \times \mathbb{C}, \quad r = 0, \ldots, p-1,$$

is unique and given by

$$q(z) = \sum_{r=0}^{p-1} \widetilde{g}(z_r^{AB}) \widetilde{L}_r^{AB}(z) = \sum_{r=0}^{p-1} g(x_r^A) e^{-2\pi i \left(c^B + \frac{\operatorname{diam} B}{2}\right) x_r^A / N} l_r^{AB}(x).$$

Using the monomial basis representation $q(z) = \sum_{s=0}^{p-1} \hat{g}_s^{AB} z^s$ with appropriate coefficients $\hat{g}_s^{AB} \in \mathbb{C}$ yields the assertion since

$$e^{2\pi i \left(c^B + \frac{\operatorname{diam} B}{2}\right) x / N} q(z(x)) = \sum_{s=0}^{p-1} \hat{g}_s^{AB} e^{2\pi i \mathfrak{z}_s^B x / N} = \mathcal{J}_p^{AB} g(x)$$

lies in $E_B^p(A)$. Furthermore, we have

$$\max_{x \in A} |\mathcal{J}_p^{AB} g(x)| = \max_{x \in A} \left| e^{2\pi i \left(c^B + \frac{\operatorname{diam} B}{2}\right) x / N} \sum_{r=0}^{p-1} g(x_r^A) e^{-2\pi i \left(c^B + \frac{\operatorname{diam} B}{2}\right) x_r^A / N} l_r^{AB}(x) \right|$$

$$\leq \|g\|_{C(A)} \max_{x \in A} \sum_{r=0}^{p-1} |l_r^{AB}(x)|.$$

Without loss of generality, let now the box B be such that $\operatorname{diam} A \operatorname{diam} B = N$. We have $x_r^A = c^A + \frac{t_r}{2} \operatorname{diam} A$. We define the mapping $y \colon A \to [-\frac{1}{2}, \frac{1}{2}]$ by

$$y(x) := \frac{1}{\operatorname{diam} A}(x - c^A) = \frac{\operatorname{diam} B}{N}(x - c^A). \tag{4.17}$$

Using

$$z(x(y)) = e^{-2\pi i \left(\frac{N}{\operatorname{diam} B} y + c^A\right) \frac{\operatorname{diam} B}{(p-1)N}} = e^{-2\pi i \frac{y}{p-1}} e^{-2\pi i c^A \frac{\operatorname{diam} B}{(p-1)N}} \quad \text{and}$$

$$z_j^A = z(x(t_j/2)) = e^{-\pi i \frac{t_j}{p-1}} e^{-2\pi i c^A \frac{\operatorname{diam} B}{(p-1)N}},$$

we obtain the normalized Lagrange functions $l_r \colon [-\frac{1}{2}, \frac{1}{2}] \to \mathbb{C}$, $r = 0, \ldots, p-1$, which are defined by

$$l_r(y) := l_r^{AB}(x(y)) = \prod_{\substack{j=0 \\ j \neq r}}^{p-1} \frac{e^{-\pi i \frac{2y}{p-1}} - e^{-\pi i \frac{t_j}{p-1}}}{e^{-\pi i \frac{t_r}{p-1}} - e^{-\pi i \frac{t_j}{p-1}}}. \tag{4.18}$$

For $a, b \in \mathbb{R}$ follows $|e^{ia} - e^{-ib}|^2 = 2 - 2\cos(a - b)$. This yields the relation

$$|l_r^{AB}(x)|^2 = |l_r(y)|^2 = \prod_{\substack{j=0 \\ j \neq r}}^{p-1} \frac{1 - \cos\left(\frac{\pi}{p-1}(2y - t_j)\right)}{1 - \cos\left(\frac{\pi}{p-1}(t_r - t_j)\right)}.$$

Since $2y - t_j \in [-2, 2]$ and $t_r - t_j \in [-2, 2]$, we apply Lemma 4.3, estimates (4.11) and (4.10), to obtain

$$\leq \prod_{\substack{j=0 \\ j \neq r}}^{p-1} \frac{2\pi^2}{\left(1 - \cos \frac{2\pi}{p-1}\right)(p-1)^2} \frac{(2y - t_j)^2}{(t_r - t_j)^2}$$

$$= K_p^2 \prod_{\substack{j=0 \\ j \neq r}}^{p-1} \left(\frac{2y - t_j}{t_r - t_j}\right)^2 = K_p^2 (L_r(2y))^2.$$

The assertion follows by Lemma 2.14, estimate (2.13), in

$$\max_{x \in A} \sum_{r=0}^{p-1} |l_r^{AB}(x)| \leq K_p \max_{x \in [-1,1]} \sum_{r=0}^{p-1} |L_r(x)| \leq K_p (1 + \frac{2}{\pi} \log p).$$

Setting $x = \frac{2\pi}{p-1}$ in Lemma 4.3 (4.12) finally yields $\lim_{p \to \infty} K_p = 1$ and $K_p \leq \frac{\pi^2}{4}$. Since $\left(\frac{2(1 - \cos x)}{x^2}\right)^{\frac{\pi}{x}} \leq 1$ for $x \in (0, \pi]$, see also the proof of Lemma 4.3, it follows $K_p \geq 1$. $\square$

We present an error estimate for the interpolation $\mathcal{J}_p^{AB} g$ of a function $g \in E_B(A)$ under the admissibility condition (4.5). Its main ingredient is the following Taylor expansion of the power function on the unit circle.

Lemma 4.5. *Let* $f_p \in C(\mathbb{R})$, $p \in \mathbb{N}$, $f_p(\alpha) := p! |\binom{\alpha}{p}| = \prod_{j=0}^{p-1} |\alpha - j|$. *It follows*

(i) $f_{p+1}(\frac{p}{2} + \alpha) = f_{p+1}(\frac{p}{2} - \alpha)$ *and*

(ii) $\frac{1}{p!} \max_{\alpha \in [0, p-1]} f_p(\alpha) \leq \frac{1}{2\pi p}$ *for* $p \geq 5$.

Proof. (i) For p even, it follows

$$f_{p+1}\left(\frac{p}{2} + \alpha\right) = \prod_{j=0}^{p} \left|\frac{p}{2} + \alpha - j\right| = \prod_{l=-\frac{p}{2}}^{\frac{p}{2}} |l + \alpha| = \prod_{l=-\frac{p}{2}}^{0} |\alpha + l| \prod_{l=1}^{\frac{p}{2}} |\alpha + l|$$

$$= |\alpha| \prod_{l=1}^{\frac{p}{2}} |\alpha - l| |\alpha + l| = |\alpha| \prod_{l=1}^{\frac{p}{2}} |\alpha - l| \prod_{l=-\frac{p}{2}}^{-1} |\alpha - l|$$

$$= f_{p+1}\left(\frac{p}{2} - \alpha\right).$$

For p odd, we have

$$f_{p+1}\left(\frac{p}{2} + \alpha\right) = \prod_{j=0}^{p} \left|\frac{p+1}{2} - \frac{1}{2} + \alpha - j\right| = \prod_{l=-\frac{p+1}{2}+1}^{\frac{p+1}{2}} \left|l - \frac{1}{2} + \alpha\right|$$

$$= \left|\alpha - \frac{1}{2}\right| \left(\prod_{l=-\frac{p+1}{2}+1}^{-1} \left|l - \frac{1}{2} + \alpha\right|\right) \left|\frac{p}{2} + \alpha\right| \prod_{l=1}^{\frac{p+1}{2}-1} \left|l - \frac{1}{2} + \alpha\right|$$

$$= \left|\alpha - \frac{1}{2}\right| \left|\frac{p}{2} + \alpha\right| \prod_{l=1}^{\frac{p-1}{2}} \left|l - \frac{1}{2} + \alpha\right| \cdot \left|-l - \frac{1}{2} + \alpha\right|$$

and finally

$$f_{p+1}\left(\frac{p}{2} - \alpha\right) = \prod_{j=0}^{p} \left|\frac{p-1}{2} + \frac{1}{2} - \alpha - j\right| = \prod_{l=-\frac{p+1}{2}}^{\frac{p-1}{2}} \left|l + \frac{1}{2} - \alpha\right|$$

$$= \prod_{l=-\frac{p+1}{2}}^{-1} \left|l + \frac{1}{2} - \alpha\right| \prod_{l=0}^{\frac{p-1}{2}} \left|l + \frac{1}{2} - \alpha\right|$$

$$= \left|-\frac{p}{2} - \alpha\right| \cdot \left|\frac{1}{2} - \alpha\right| \left(\prod_{l=-\frac{p-1}{2}}^{-1} \left|l + \frac{1}{2} - \alpha\right|\right) \prod_{l=1}^{\frac{p-1}{2}} \left|l + \frac{1}{2} - \alpha\right|$$

$$= \left|\frac{p}{2} + \alpha\right| \cdot \left|\alpha - \frac{1}{2}\right| \prod_{l=1}^{\frac{p-1}{2}} \left|l + \frac{1}{2} - \alpha\right| \left|-l + \frac{1}{2} - \alpha\right|$$

$$= \left|\frac{p}{2} + \alpha\right| \cdot \left|\alpha - \frac{1}{2}\right| \prod_{l=1}^{\frac{p-1}{2}} \left|-l - \frac{1}{2} + \alpha\right| \left|l - \frac{1}{2} + \alpha\right|$$

$$= f_{p+1}\left(\frac{p}{2} + \alpha\right).$$

(ii) We prove by induction. Since the function $\frac{1}{5!}f_5$ takes its maximum in

$$\alpha_0 = \frac{1}{10}\left(20 + \sqrt{10(15 - \sqrt{145})}\right) \approx 0.355567 \in [0, p-1],$$

it follows

$$\frac{1}{p!} \max_{\alpha \in [0, p-1]} f_p(\alpha) \approx 0.0302619 \leq \frac{1}{2\pi \cdot 5} \quad \text{for } p = 5.$$

Furthermore, we have $f_{p+1}(\frac{p}{2} + \alpha) = f_{p+1}(\frac{p}{2} - \alpha)$ and $[0, \frac{p}{2}] \subset [0, p-1]$ for $p \geq 2$. It follows for $p+1$,

$$\frac{1}{(p+1)!} \max_{\alpha \in [0,p]} f_{p+1}(\alpha) = \frac{1}{p+1} \max_{\alpha \in [0,\frac{p}{2}]} \frac{1}{p!} |\alpha - p| \prod_{j=0}^{p-1} |\alpha - j|$$

$$\leq \frac{p}{p+1} \max_{\alpha \in [0,\frac{p}{2}]} \frac{1}{p!} \prod_{j=0}^{p-1} |\alpha - j| \leq \frac{p}{p+1} \frac{1}{2\pi p} = \frac{1}{2\pi(p+1)}.$$

$\square$

Definition 4.6. *We define the principal value of the complex logarithm*

$$\text{Log}: \mathbb{C} \setminus (-\infty, 0] \to \mathbb{C},$$

such that

$$\text{Log}(z) := \log|z| + \mathrm{i}\varphi, \quad \text{with } z = |z|\mathrm{e}^{\mathrm{i}\varphi}, \quad \varphi \in (-\pi, \pi)$$

and for $\alpha \in \mathbb{C}$, the principal value of complex power $f \colon \mathbb{C} \setminus (-\infty, 0] \to \mathbb{C}$ by

$$f(z) = z^\alpha := \mathrm{e}^{\alpha \text{Log} z}.$$

Lemma 4.7. *Let $p \in \mathbb{N}$, $p \geq 5$, $\Gamma_p := \{z \in \mathbb{C} : z = \mathrm{e}^{-2\pi \mathrm{i} x/(p-1)}, -\frac{1}{2} \leq x \leq \frac{1}{2}\}$, and $\alpha \in [0, p-1]$, then we have for $z \in \Gamma_p$ the estimate*

$$\left| z^\alpha - \sum_{k=0}^{p-1} \binom{\alpha}{k} (z-1)^k \right| \leq c_p$$

with the constant

$$c_p := \frac{1}{\pi p} \left(\frac{\pi}{p-1} \right)^p. \tag{4.19}$$

Proof. The function $g : \mathbb{C} \setminus (-\infty, 0] \to \mathbb{C}$, $g(z) := z^\alpha$, is holomorphic and can be represented by its Taylor series at $z = 1$ in Γ_p for $p \geq 5$. Using the estimate (4.10), yields

$$|z-1|^2 = 2 \left(1 - \cos \frac{2\pi x}{p-1} \right) \leq \left(\frac{2\pi x}{p-1} \right)^2 \leq \left(\frac{\pi}{p-1} \right)^2 < 1 \tag{4.20}$$

and we finally follow the ideas in the proof of [52, Theorem 1]. We set $a_k := (-1)^k \binom{\alpha}{k}$. Due to $a_{k+1} = \frac{k-\alpha}{k+1} a_k$, the coefficients a_k have the same sign and this yields $|a_{k+1}| \leq |a_k|$ for $\alpha \leq p \leq k$. Due to $|z-1| < 1$, the series $\sum_{k=0}^{\infty} (z-1)^k$ is absolutely convergent. Since the binomial series $(1+x)^\alpha = \sum_{k=0}^{\infty} \binom{\alpha}{k} x^k$ converges absolutely and uniformly for $\alpha \geq 0$ and $x \in [-1, 1]$, it follows that the series $\sum_{k=0}^{\infty} \binom{\alpha}{k}$ is absolutely convergent too. Now, we can use Abel summation and obtain

$$
\begin{aligned}
\left| z^\alpha - \sum_{k=0}^{p-1} \binom{\alpha}{k}(z-1)^k \right| &= \left| \sum_{k=p}^{\infty} \binom{\alpha}{k}(z-1)^k \right| = \left| \sum_{k=p}^{\infty} a_k(1-z)^k \right| \\
&= \left| \sum_{k=p}^{\infty} (a_k - a_{k+1}) \sum_{l=p}^{k} (1-z)^l \right| \\
&= \left| \sum_{k=p}^{\infty} (a_k - a_{k+1}) \frac{(1-z)^p - (1-z)^{k+1}}{1 - (1-z)} \right| \\
&\leq |z-1|^p \sum_{k=p}^{\infty} |a_k - a_{k+1}| \frac{|1 - (1-z)^{k+1-p}|}{|z|}. \tag{4.21}
\end{aligned}
$$

Due to $z \in \Gamma_p$, we finally have

$$
\begin{aligned}
&\leq 2|z-1|^p \sum_{k=p}^{\infty} |\operatorname{sign}(a_k)(|a_k| - |a_{k+1}|)| \\
&= 2|z-1|^p |a_p| = 2|z-1|^p \left| (-1)^p \binom{\alpha}{p} \right|.
\end{aligned}
$$

The assertion follows by applying the estimate (4.20) and the previous Lemma 4.5.

$\square$

Theorem 4.8. *Let $p, N \in \mathbb{N}$, $p \geq 5$, two boxes $A, B \subset \mathbb{R}$ be admissible in the sense*

$$\operatorname{diam} A \operatorname{diam} B \leq N,$$

and $g \in E_B(A)$, $g(x) := \sum_{k=1}^{M_2} \hat{g}_k e^{2\pi i \xi_k x / N}$, $\hat{g}_k \in \mathbb{C}$, $k = 1, \ldots, M_2$, then we have the error estimate

$$\|g - \mathcal{J}_p^{AB} g\|_{C(A)} \leq (1 + C_p) \cdot c_p \cdot \|\hat{\boldsymbol{g}}\|_1,$$

where the constants are given by (4.15) and (4.19), respectively.

Proof. Without loss of generality, let the box B be such that $\operatorname{diam} A \operatorname{diam} B = N$. We shift and dilate $A \mapsto [-\frac{1}{2}, \frac{1}{2}]$ and $\widetilde{\xi} \colon B \to [-\frac{N}{2}, \frac{N}{2}]$ via the map in (4.17) and

$$\widetilde{\xi} := \widetilde{\xi}(\xi) = \frac{N}{\operatorname{diam} B}(\xi - c^B) = \operatorname{diam} A(\xi - c^B), \tag{4.22}$$

respectively. Inserting the maps for x and ξ in g yields with $\widetilde{g} \colon [-\frac{1}{2}, \frac{1}{2}] \to \mathbb{C}$,

$$\widetilde{g}(y) := \sum_{k=1}^{M_2} \hat{g}_k e^{2\pi i \widetilde{\xi}_k c^A / (N \operatorname{diam} A)} e^{2\pi i \widetilde{\xi}_k y / N}, \tag{4.23}$$

the representation

$$g(x) = e^{2\pi i c^B x / N} \sum_{k=1}^{M_2} \hat{g}_k e^{2\pi i \widetilde{\xi}_k x / (N \operatorname{diam} A)} = e^{2\pi i c^B x / N} \widetilde{g}(y).$$

Moreover, the interpolation operator fulfills a similar shift property

$$\mathcal{J}_p^{AB} g(x) = e^{2\pi i c^B x / N} \mathcal{J}_p \widetilde{g}(y), \quad \mathcal{J}_p := \mathcal{J}_p^{[-\frac{1}{2}, \frac{1}{2}][-\frac{N}{2}, \frac{N}{2}]}.$$

Now, we set $z \colon [-\frac{1}{2}, \frac{1}{2}] \to \Gamma_p$, $z := z(y) = e^{-2\pi i y / (p-1)}$. Applying z to the interpolation operator yields

$$\mathcal{J}_p \widetilde{g}(y) = z^{-\frac{p-1}{2}} \sum_{k=1}^{M_2} \widetilde{g}(t_s/2)(z(t_s/2))^{\frac{p-1}{2}} \widetilde{L}_s^{[-\frac{1}{2}, \frac{1}{2}][-\frac{N}{2}, \frac{N}{2}]}(z)$$

and setting $h \colon \Gamma_p \to \mathbb{C}$,

$$h(z) := z^{\frac{p-1}{2}} \sum_{k=1}^{M_2} \hat{h}_k z^{-(p-1)\widetilde{\xi}_k / N}, \quad \hat{h}_k := \hat{g}_k e^{2\pi i \widetilde{\xi}_k c^A / (N \operatorname{diam} A)}, \tag{4.24}$$

leads to

$$\widetilde{g}(y) = z^{-\frac{p-1}{2}} h(z),$$
$$\mathcal{J}_p \widetilde{g}(y) = z^{-\frac{p-1}{2}} \mathcal{I}_p h(z), \quad \mathcal{I}_p := \mathcal{I}_p^{[-\frac{1}{2}, \frac{1}{2}][-\frac{N}{2}, \frac{N}{2}]}.$$

Both mappings leave the error unchanged, i.e.,

$$\max_{x \in A} |g(x) - \mathcal{J}_p^{AB} g(x)| = \max_{y \in [-\frac{1}{2}, \frac{1}{2}]} |\tilde{g}(y) - \mathcal{J}_p \tilde{g}(y)| = \max_{z \in \Gamma_p} |h(z) - \mathcal{I}_p h(z)|.$$

We rewrite $h(z) = \sum_{k=1}^{M_2} \hat{h}_k z^{\eta_k}$ with $\eta_k := (p-1)(\frac{1}{2} - \frac{\tilde{\xi}_k}{N}) \in [0, p-1]$, and approximate h by its truncated Taylor series $h_p \in \Pi_{p-1}(\mathbb{C})$ at $z = 1$,

$$h_p(z) := \sum_{k=1}^{M_2} \hat{h}_k \sum_{r=0}^{p-1} \binom{\eta_k}{r} (z-1)^r. \tag{4.25}$$

We conclude from the reproduction of polynomials, $\mathcal{I}_p h_p = h_p$, the boundedness of the operator norm, cf. proof of Lemma 4.4,

$$\|\mathcal{I}_p\| := \sup_{\substack{h \in C(\Gamma_p) \\ \|h\|_{C(\Gamma_p)} = 1}} \|\mathcal{I}_p h\|_{C(\Gamma_p)} = \|\mathcal{J}_p^{AB}\| \le C_p, \tag{4.26}$$

and Lemma 4.7 the assertion by

$$\max_{z \in \Gamma_p} |h(z) - \mathcal{I}_p h(z)| \le \max_{z \in \Gamma_p} |h(z) - h_p(z)| + \max_{z \in \Gamma_p} |h_p(z) - \mathcal{I}_p h(z)|$$

$$\le (1 + \|\mathcal{I}_p\|) \max_{z \in \Gamma_p} |h(z) - h_p(z)|$$

$$\le (1 + C_p) c_p \sum_{k=1}^{M_2} |\hat{g}_k|.$$

□

Remark 4.9. *Via the mapping $A \mapsto [-\frac{1}{2}, \frac{1}{2}] \mapsto \Gamma_p$, see Figure 4.1, we rewrite the trigonometric interpolation operator $\mathcal{J}_p^{AB}$ as an polynomial interpolation operator $\mathcal{I}_p$ in the complex plane by*

$$\mathcal{J}_p^{AB} g(x) = \mathrm{e}^{2\pi \mathrm{i} c^B x / N} z^{-\frac{p-1}{2}} \mathcal{I}_p h(z),$$

with z and h defined in the proof of the previous theorem. This result gives us a connection between the trigonometric interpolation in Definition 4.1 and the complex polynomial interpolation in Definition 4.2 with an appropriate function h.

Remark 4.10. *Let $\tilde{B} := [-\frac{\mathrm{diam}\, B}{2}, \frac{\mathrm{diam}\, B}{2}]$ be a centrally located box around the origin and $g \in E_{\tilde{B}}(A)$. If a Bernstein inequality holds true on a bounded domain of the form*

$$\|g'\|_{C(A)} \le C \frac{\mathrm{diam}\, B}{N} \|g\|_{C(A)} \tag{4.27}$$

with some constant $C > 0$, the estimate in Theorem 4.8 can be reduced to

$$\|g - \mathcal{J}_p^{AB} g\|_{C(A)} \le D_p \|g\|_{C(A)}$$

Figure 4.1: The map $A \mapsto [-\frac{1}{2}, \frac{1}{2}] \mapsto \Gamma_p$.

with a constant $D_p := \frac{2C^p}{4^{p-1}p! - C^p}$, see [41, Satz 26]. Moreover, the operatornorm can be bounded by

$$\|\mathcal{J}_p^{AB}\| \le (D_p + 1),$$

see [41, Folgerung 27]. Here, the error can be estimated against the maximum norm of the function g, whereas the error in Theorem 4.8 is estimated against the 1-norm of the coefficients of the function g. The estimate against the maximum norm is a stronger statement. But note that for $g \in E_{\tilde{B}}(A)$ a Bernstein inequality [61, p. 104] has only been proven for $A = \mathbb{R}$. This proof does not allow a restriction on a bounded domain. The main consequence of Theorem 4.8 is an increase of the operator norm from a constant to $\log p$ depending on the approximation rank p, which also has to be expected in the context of interpolation in Chebyshev nodes.

Remark 4.11. *In [58], the extremal points $t_j^{\max}$, $j = 0, \ldots, p-1$, of the Chebyshev polynomials T_{p-1} were used in space and frequency domain.*
In contrast, we use the Chebyshev nodes t_j, $j = 0, \ldots, p-1$, in the spatial domain and equispaced nodes β_s, $s = 0, \ldots, p-1$, in the frequency domain. This choice of nodes assures an explicit representation of the interpolating function $\mathcal{J}_p^{AB} g$ by the trigonometric Lagrange functions l_r^{AB} and finally a local error analysis for two admissible boxes A, B as in the previous Theorem. An explicit representation for the interpolation in both variables is given in Definition 4.12. From the error estimate given in the previous Theorem 4.8, we deduce a local error estimate for interpolating in both variables in Chebyshev nodes, see Corollary 4.13. An estimate for the original variant, interpolating in both variables in the extremal points can be proven analogously.

We define the interpolation in both variables of the Fourier kernel.

Definition 4.12. *Let $A, B \subset [0, \infty)$, $\kappa \colon A \times B \to \mathbb{C}$, $\kappa(x, \xi) := e^{2\pi i x \xi / N}$ and $p \in \mathbb{N}$, $p \ge 2$. We denote by $\mathcal{I}$ the identity, then we define the interpolations in one variable via*

$$(\mathcal{J}_p^{AB} \otimes \mathcal{I})\kappa(x, \xi) := e^{2\pi i \left(c^B + \frac{\text{diam } B}{2}\right) x / N} \sum_{r=0}^{p-1} \kappa(x_r^A, \xi) e^{-2\pi i \left(c^B + \frac{\text{diam } B}{2}\right) x_r^A / N} l_r^{AB}(x),$$

$$(\mathcal{I} \otimes \mathcal{J}_p^{BA})\kappa(x, \xi) := e^{2\pi i \left(c^A + \frac{\text{diam } A}{2}\right) \xi / N} \sum_{s=0}^{p-1} \kappa(x, \xi_s^B) e^{-2\pi i \left(c^A + \frac{\text{diam } A}{2}\right) \xi_s^A / N} l_r^{BA}(\xi),$$

and in two variables via

$$(\mathcal{J}_p^{AB} \otimes \mathcal{J}_p^{BA})\kappa(y,\xi) := d^{AB}(x,\xi) \sum_{r=0}^{p-1} \sum_{s=0}^{q-1} l_r^A(x) d^{AB}(-x_r^A, -\xi_s^B) \kappa(x_r^A, \xi_s^B) l_s^B(\xi),$$

where

$$d^{AB}(x,\xi) := e^{2\pi i \left(\left(c^B + \frac{\text{diam } B}{2} \right) x + \left(c^A + \frac{\text{diam } A}{2} \right) \xi \right)/N}.$$

We have

$$(\mathcal{J}_p^{AB} \otimes \mathcal{I})(\mathcal{I} \otimes \mathcal{J}_p^{BA})\kappa(x,\xi) = (\mathcal{J}_p^{AB} \otimes \mathcal{J}_p^{BA})\kappa(y,\xi).$$

Corollary 4.13. *Under the conditions and definitions in Theorem 4.8 and for $p \in \mathbb{N}$, $p \geq 5$, we have*

$$\left\| \kappa - (\mathcal{J}_p^{AB} \otimes \mathcal{J}_p^{BA})\kappa \right\|_{C(A \times B)} \leq \frac{961}{900} \cdot \left(\frac{\pi}{p-1} \right)^p.$$

Proof. Applying Theorem 4.8 yields

$$\left\| \kappa - \left(\mathcal{J}_p^{AB} \otimes \mathcal{I} \right) \kappa \right\|_{C(A \times B)} \leq (C_p + 1)c_p.$$

Factorizing the interpolation operator and applying Theorem 4.8 yields

$$\begin{aligned}
\left\| \kappa - (\mathcal{J}_p^{AB} \otimes \mathcal{J}_p^{BA})\kappa \right\|_{C(A \times B)} &\leq \left\| \kappa - \left(\mathcal{J}_p^{AB} \otimes \mathcal{I} \right) \kappa \right\|_{C(A \times B)} \\
&\quad + \left\| \mathcal{J}_p^{AB} \otimes \mathcal{I} \right\| \left\| \kappa - \left(\mathcal{I} \otimes \mathcal{J}_p^{BA} \right) \kappa \right\|_{C(A \times B)} \\
&\leq (1 + C_p)c_p + C_p(1 + C_p)c_p = (1 + C_p)^2 c_p.
\end{aligned}$$

Let $f_1, f_2 : [5, \infty) \to \mathbb{R}$, $f_1(p) := 1 + \frac{\pi^2}{4} + \frac{\pi}{2} \log p$ and $f_2(p) := \frac{31}{30}\sqrt{\pi p}$, be given. First of all, we prove

$$1 + C_p \leq f_1(p) < f_2(p). \tag{4.28}$$

The first inequality follows from the definition of C_p, see Equation (4.15) and from $K_p \leq \frac{\pi^2}{4}$, see Equation (4.16). Moreover, we have $f_1(5) < f_2(5)$. Due to $p \geq 5 \geq \pi$, it follows $\frac{1}{p} \leq \frac{1}{\sqrt{\pi p}}$ and finally

$$f_1'(p) = \frac{\pi}{2p} \leq \frac{\pi}{2\sqrt{\pi p}} = \frac{\sqrt{\pi}}{2\sqrt{p}} < f_2'(p).$$

Since f_1 and f_2 are continuously differentiable, Inequality (4.28) follows and this implies

$$(1 + C_p)^2 c_p \leq \left(\frac{31}{30} \right)^2 \left(\frac{\pi}{p-1} \right)^p.$$

$\square$

Summarizing, the interpolation in two variables by Chebyshev nodes has nearly the same convergence behavior as interpolation in the first variable and using equispaced frequencies.

Regarding the optimality of these error estimates, the following is known.

Remark 4.14. *The frequency and time limited Fourier transform*

$$\mathcal{F}_0 \colon L^2[-\tfrac{1}{2}, \tfrac{1}{2}] \to L^2[-\tfrac{1}{2}, \tfrac{1}{2}], \quad f \mapsto g(\xi) = \int_{-\frac{1}{2}}^{\frac{1}{2}} f(x) \mathrm{e}^{-2\pi \mathrm{i}\xi x} \mathrm{d}x,$$

fulfills

$$\mathcal{F}_0^* \mathcal{F}_0 f(x) = \int_{-\frac{1}{2}}^{\frac{1}{2}} \frac{\sin \pi(x - y)}{x - y} f(y) \mathrm{d}y$$

and thus the singular values of $\mathcal{F}_0$ are

$$\sigma_p \approx C \left(\frac{\pi}{8}\right)^p \frac{1}{p!} \approx C' \left(\frac{1.06}{p}\right)^p$$

see [57, Lemma 1]. In particular, this best possible approximation rate would be guaranteed when approximating by prolate spheroidal wave functions, cf. [38], or adaptive cross approximation with total pivoting, cf. [55]. Corollary 4.13 implies $\sigma_p \leq \frac{1}{2}(\pi/(p-1))^p$ and this slower rate cannot be improved by the current proof technique in Lemma 4.7. However, note that the numerical observed rate of the interpolatory approximation matches the optimal one, see Figure 4.8(a).

4.1.2 Butterfly algorithm

We present the whole approximation scheme and the corresponding algorithm to compute the sums (4.1).

4.1.2.1 Dyadic decomposition and approximation scheme

For two admissible boxes, see Definition 4.1, Theorem 4.8 ensures a low rank approximation of the Fourier kernel $\kappa(x, \xi) = \mathrm{e}^{2\pi \mathrm{i}x\xi/N}$ via interpolation. We declare admissible pairs $(A, B) \subset X \times \Omega$ for a convenient decomposition of the domains X and Ω. Since $X = \Omega = [0, 2^L]$ a simple way is to decompose the domains dyadically, see Figure 4.2.

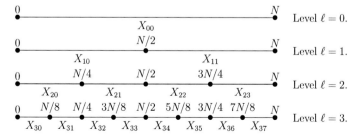

Figure 4.2: Dyadic decomposition of the spatial domain X.

The dyadic decomposition is given by

$$
\begin{aligned}
X_{\ell,m} &:= \left[N/2^\ell m, N/2^\ell (m+1) \right) && \text{for } m = 0, \dots, 2^\ell - 1, \\
\Omega_{\ell,n} &:= \left[N/2^\ell n, N/2^\ell (n+1) \right) && \text{for } n = 0, \dots, 2^\ell - 1,
\end{aligned}
\tag{4.29}
$$

for $\ell = 0, \dots, L$, where the level of the butterfly scheme is denoted by ℓ and the locations of the boxes by m and n, respectively. Moreover, note that we always include the point N in the rightmost sets $X_{\ell,2^\ell-1}$ and $\Omega_{\ell,2^\ell-1}$, $\ell = 0, \dots, L$. The pairs $(X_{\ell,m}, \Omega_{L-\ell,n})$ for $m = 0, \dots, 2^\ell - 1$ and $n = 0, \dots, 2^{L-\ell} - 1$ are admissible, which is due to diam $X_{\ell,m}$ diam $\Omega_{L-\ell,n} = \frac{N}{2^\ell} \frac{N}{2^{L-\ell}} = N$. All admissible pairs in the sense of Inequality (4.5) of the dyadic decomposition are illustrated for $X = \Omega = [0,8]$ in Figure 4.3. The result of the dyadic decomposition of X and Ω is the formation of two

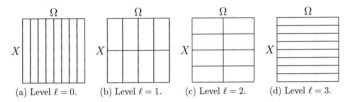

(a) Level $\ell = 0$.   (b) Level $\ell = 1$.   (c) Level $\ell = 2$.   (d) Level $\ell = 3$.

Figure 4.3: Admissible pairs of $X = \Omega = [0,8]$.

trees $\mathcal{T}_X$ and $\mathcal{T}_\Omega$, where $X_{00} = X$ is the root node of $\mathcal{T}_X$ and $\Omega_{00} = \Omega$ is the root node of the tree $\mathcal{T}_\Omega$. Let $X_{\ell,m} \cap \widetilde{X} \neq \emptyset$, $m \in \{0, \dots, 2^\ell - 1\}$, be a node of the tree $\mathcal{T}_X$. Then $X_{\ell+1,2m}$ is connected by an edge to $X_{\ell,m}$, if $X_{\ell+1,2m} \cap \widetilde{X} \neq \emptyset$, analogously for $X_{\ell+1,2m+1}$. The tree $\mathcal{T}_\Omega$ is build up analogously. The complete trees are illustrated in Figure 4.4(a), 4.4(b), where the tree $\mathcal{T}_\Omega$ is represented reverted from the major level L (the leaf nodes) to the least level 0 (the root node). We define the butterfly graph as the product of the two trees $\mathcal{T}_X$ and $\mathcal{T}_\Omega$, in which all admissible pairs $(X_{\ell,m}, \Omega_{L-\ell,n})$ are nodes in the butterfly graph. The nodes in the butterfly graph are given as combinations of nodes in the ℓ-th level of the tree $\mathcal{T}_X$ and the $(L - \ell)$-th level in the tree $\mathcal{T}_\Omega$. An edge $\left\{ (X_{\ell,m}, \Omega_{L-\ell,n}), (X_{\ell+1,m'}, \Omega_{L-(\ell+1),n'}) \right\}$ is set, if and only if $(X_{\ell,m}, X_{\ell,m'})$ and $(\Omega_{L-\ell,n}, \Omega_{L-\ell,n'})$ are edges in the trees $\mathcal{T}_X$ and $\mathcal{T}_\Omega$, respectively. The complete butterfly graph has $L + 1$ levels and $2^L = N$ nodes in each level, see Figure 4.4(c). Note that neither the trees nor the butterfly graph have to be complete. This depends on the position of the sampling nodes $\widetilde{X}$ and $\widetilde{\Omega}$. The trees were realized as a recursive data structure, see Algorithm 2. The data structure includes the level ℓ and the parameter m of the box $X_{\ell,m}$ as well as the set of sons S, if they exist, as a set of trees. If $X_{\ell,m}$ is a leaf node in the tree, then the index set I of nodes x_j, such that $x_j \in X_{\ell,m}$, is stored. To generate the tree $\mathcal{T}_X$ for our setting, we have to call the constructor via

$$
\mathcal{T}_X := \text{TREE}(\widetilde{X}, (1, \dots, M_1), 0, 0).
$$

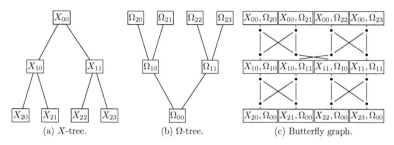

Figure 4.4: Trees and butterfly graph for $N = 4$.

Algorithm 2 Constructor tree in 1D

Input:

$N = 2^L,\ L \in \mathbb{N}$ ▷ bandwidth

$M_1 \in \mathbb{N}$ ▷ number of sampling nodes

$\widetilde{X} = \{x_1, \ldots, x_{M_1}\} \subset X$ ▷ sampling nodes

$I \subset \{1, \ldots, M_1\}$ ▷ index set

$\ell \in \{0, 1, \ldots, L\}$ ▷ level

$m \in \{0, 1, \ldots, 2^\ell - 1\}$ ▷ position parameter

    function TREE($\widetilde{X}, I, N, \ell, m$)

        $\mathcal{T}.\ell := \ell$

        $\mathcal{T}.m := m$

        if $2^{\mathcal{T}.\ell} = N$ **or** $I = \emptyset$ **then**

            $\mathcal{T}.I := I$

        else

            for $r = 0, 1$ **do**

                $I_r := \{j \in I : x_j \in X_{\ell+1, 2m+r}\}$

                if $I_r \neq \emptyset$ **then**

                    $s := \text{TREE}(\widetilde{X}, I_r, \ell + 1, 2m + r)$ ▷ son of data type TREE

                    $\mathcal{T}.S := \{\mathcal{T}.S, s\}$ ▷ set of sons

                end if

            end for

        end if

    end function

Output:

$\mathcal{T}$ ▷ tree

Algorithm 3 Butterfly sparse fast Fourier transform in dimension one (BSFFT)

Input:

$L, M_1, M_2, p \in \mathbb{N}, \, p \geq 2, \, N = 2^L,$

$\hat{f}_k \in \mathbb{C}, \, k = 1, \ldots, M_2,$

$\Omega = \{\xi_k \in X : k = 1, \ldots, M_2\},$

$\widetilde{X} = \{x_j \in \Omega : j = 1, \ldots, M_1\}.$

---

1. Decompose the space and frequency domain X and Ω dyadically, see Equation (4.29).

2. Define the functions $f^{\Omega_{L,n}} \colon X \to \mathbb{C}$ and the corresponding interpolants $f^{X_{0,0}\Omega_{L,n}}$ by

   for $n = 0, \ldots, 2^L - 1$ **do**

   $$f^{\Omega_{L,n}}(x) := \sum_{\xi_j \in \Omega_{L,n} \cap \widetilde{\Omega}} \hat{f}_j e^{2\pi i \xi_j x / N}, \tag{4.30}$$

   $$f^{X_{0,0}\Omega_{L,n}} := \mathcal{J}_p^{X_{0,0}\Omega_{L,n}} f^{\Omega_{L,n}}.$$

   end for

3. Compute for the levels $\ell = 1, \ldots, L$ the recursively defined interpolating functions $f^{X_{\ell,m}\Omega_{L-\ell,n}} \colon X_{\ell,m} \to \mathbb{C}$.

   for $\ell = 1, \ldots, L$ **do**
       for $m = 0, \ldots, 2^\ell - 1$ and $n = 0, \ldots, 2^{L-\ell} - 1$ **do**

   $$f^{X_{\ell,m}\Omega_{L-\ell,n}} := \mathcal{J}_p^{X_{\ell,m}\Omega_{L-\ell,n}} \left[f^{X_{\ell-1,\lfloor m/2 \rfloor}\Omega_{L-\ell+1,2n}} + f^{X_{\ell-1,\lfloor m/2 \rfloor}\Omega_{L-\ell+1,2n+1}} \right]$$

       end for
   end for

4. The final approximation $\widetilde{f}$ of $f \in E_\Omega(X)$ is a composed function of approximating functions from the L-th level and given by

   for $j = 1, \ldots, M_1$ **do**

   $$m := \lfloor x_j \rfloor$$
   $$\widetilde{f}(x_j) := f^{X_{L,m}\Omega_{0,0}}(x_j).$$

   end for

Output:

Approximate function values $\widetilde{f}(x_j), \, j = 1, \ldots, M_1.$

4.1.2.2 Algorithm

We quote Algorithm 3, which computes the function values $\widetilde{f}(x_j)$ as an approximation to the function values $f(x_j)$ for $j = 1, \ldots, M_1$, $f \in E_\Omega(X)$, by traversing the butterfly scheme, see Figure 4.4(c). We denote Algorithm 3 as butterfly sparse fast Fourier transform, short BSFFT, even though the one-dimensional set of nodes has not to be sparse. Note that the approximate function $\widetilde{f} \colon X \to \mathbb{C}$ is defined as a decomposed function by $\widetilde{f}|_{X_{L,m}} = f^{X_{L,m}\Omega_{0,0}} \in E_\Omega(X_{L,m})$ for $m = 0, \ldots, 2^L - 1$, see Algorithm 3 Step 3.

4.1.3 Realization

In this section, we discuss how to compute the approximate function $\widetilde{f}$ in Algorithm 3. The local approximation by means of the interpolation operator needs to be realized using a basis for the ansatz space $E_B^p(A)$. Subsequently, we discuss a variant of the original approach [58] which uses a monomial type basis and a new variant which relies on a Lagrange type basis. While both approaches take approximately the same amount of computation, the latter is much more stable. Moreover, note that the boxes of the dyadic decomposition of X and Ω fulfill the relations $X_{\ell,m} \subset X_{\ell-1,\lfloor m/2 \rfloor}$ and $\Omega_{L-\ell,n} = \Omega_{L-\ell+1,2n} \dot\cup \Omega_{L-\ell+1,2n+1}$. To shorten notation let $A := X_{L,m}$ and $B := \Omega_{L-\ell,n}$ for $\ell \in \{1, \ldots, L\}$, $m \in \{0, \ldots, 2^\ell - 1\}$, and $n \in \{0, \ldots, 2^{L-\ell} - 1\}$. In Step 3, we have to compute the approximation $f^{AB} \in E_B(A)$,

$$f^{AB} := \mathcal{J}_p^{AB} \sum_{S \in S_B} f^{PS}, \tag{4.31}$$

where

$$S_B := \left\{ \left[c^B - \frac{\operatorname{diam} B}{2}, c^B \right], \left[c^B, c^B + \frac{\operatorname{diam} B}{2} \right] \right\},$$

denotes the set of sons, at most two, of the set of frequencies

$$B = \left[c^B - \frac{\operatorname{diam} B}{2}, c^B + \frac{\operatorname{diam} B}{2} \right]$$

and the interpolation error is small in each of the spatial sets, again at most two,

$$A = \left[c^A - \frac{\operatorname{diam} A}{2}, c^A + \frac{\operatorname{diam} A}{2} \right],$$

which are subsets of their father

$$P := \left[c^P - \frac{\operatorname{diam} P}{2}, c^P + \frac{\operatorname{diam} P}{2} \right].$$

Subsequently, we rely on the admissibility condition $\operatorname{diam} A \operatorname{diam} B = N$ and on the dyadic decomposition which results in $\operatorname{diam} P \operatorname{diam} B = 2N$, see Figure 4.5 for an illustration of the sets.

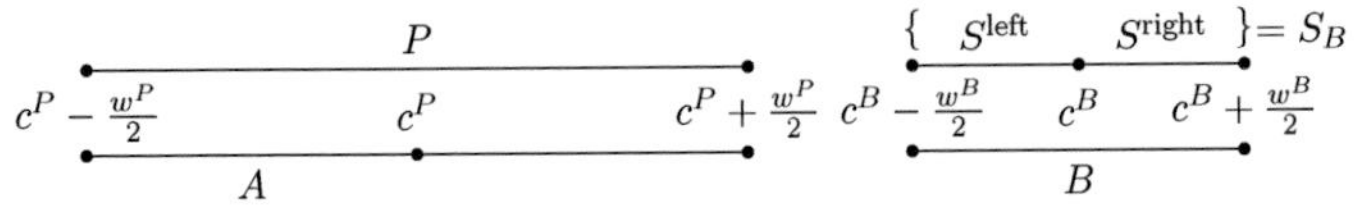

Figure 4.5: Illustration of the spatial set $A \subset P$ and frequency set $B \supset S_B$.

4.1.3.1 Monomial type basis

Let $g \in E_B(A)$. The interpolating function $\mathcal{J}_p^{AB} g$ in Definition 4.1 has a unique solution in $E_B^p(A)$ and can be represented in a monomial type basis,

$$\mathcal{J}_p^{AB} g(x) = \sum_{k=0}^{p-1} \hat{g}_s^{AB} e^{2\pi i \xi_s^B x/N}$$

where the coefficient vector $\hat{\mathbf{g}}^{AB} = (\hat{g}_s^{AB})_{s=0}^{p-1} \in \mathbb{C}^p$ is the solution of the interpolation problem

$$\mathbf{g}^{AB} = \mathbf{M}^{AB} \hat{\mathbf{g}}^{AB}, \quad \text{where} \quad \mathbf{g}^{AB} := (g(x_r^A))_{r=0}^{p-1}, \tag{4.32}$$

with the interpolation matrix

$$\mathbf{M}^{AB} := \left(e^{2\pi i \xi_s^B x_r^A/N} \right)_{r,s=0}^{p-1}, \tag{4.33}$$

equally spaced nodes $\xi_s^B \in B$, and Chebyshev nodes $x_r^A \in A$, cf. Equation (4.3) and Equation (2.5). We have

$$e^{2\pi i \xi_s^B x_r^A/N} = e^{2\pi i (c^B + \beta_s \, \text{diam} \, B)(c^A + \frac{t_r^A}{2} \, \text{diam} \, A)/N} = e^{2\pi i c^B x_r^A/N} e^{\pi i \beta_s t_r} e^{2\pi i c^A \beta_s \, \text{diam} \, B/N}$$

and using the two diagonal matrices

$$\mathbf{D}_1^{AB} := \text{diag} \left(e^{2\pi i x_r^A c^B/N} \right)_{r=0}^{p-1}, \quad \mathbf{D}_2^{AB} := \text{diag} \left(e^{2\pi i c^A \beta_s \, \text{diam} \, B/N} \right)_{s=0}^{p-1},$$

the matrix $\mathbf{M}^{AB} \in \mathbb{C}^{p \times p}$ can be factorized as

$$\mathbf{M}^{AB} = \mathbf{D}_1^{AB} \mathbf{G} \mathbf{D}_2^{AB}, \quad \mathbf{G} := \left(e^{\pi i t_r \beta_s} \right)_{r,s=0}^{p-1}, \tag{4.34}$$

where the matrix $\mathbf{G} \in \mathbb{C}^{p \times p}$ is independent of A and B. Applied to Equation (4.31), this yields

$$\hat{\mathbf{f}}^{AB} = \left(\mathbf{D}_2^{AB} \right)^{-1} \mathbf{G}^{-1} \left(\mathbf{D}_1^{AB} \right)^{-1} \mathbf{f}^{AB}, \tag{4.35}$$

where $\mathbf{f}^{AB} = \left(\sum_{S \in S_B} f^{PS}(x_r^A) \right)_{r=0}^{p-1}$. Given the coefficients $\hat{\mathbf{f}}^{PS} = (\hat{f}_s^{PS})_{s=0}^{p-1} \in \mathbb{C}^p$ in $f^{PS}(x) = \sum_{s=0}^{p-1} \hat{f}_s^{PS} e^{2\pi i \xi_s^S x/N}$, we compute

$$\mathbf{f}^{AB} = \sum_{S \in S_B} \mathbf{N}^{AS} \hat{\mathbf{f}}^{PS}, \quad \mathbf{N}^{AS} := \left(e^{2\pi i \xi_s^S x_r^A/N} \right)_{r,s=0}^{p-1}.$$

Since diam $S = \frac{\text{diam } B}{2}$ and

$$e^{2\pi i \xi_s^S x_r^A / N} = e^{2\pi i (c^S + \beta_s \, \text{diam } S)(c^A + \frac{t_r}{2} \, \text{diam } A)/N} = e^{2\pi i c^S x_r^A / N} e^{\frac{\pi}{2} i \beta_s t_r} e^{2\pi i c^A \beta_s \, \text{diam } S / N}$$

we have with the two diagonal matrices

$$\mathbf{E}_1^{AS} := \text{diag}\left(e^{2\pi i x_r^A c^S / N}\right)_{r=0}^{p-1}, \quad \mathbf{E}_2^{AS} := \text{diag}\left(e^{2\pi i c^A \beta_s \, \text{diam } S / N}\right)_{s=0}^{p-1},$$

the factorization

$$\mathbf{N}^{AS} = \mathbf{E}_1^{AS} \mathbf{H} \mathbf{E}_2^{AS}, \quad \mathbf{H} := \left(e^{\frac{\pi}{2} i t_r \beta_s}\right)_{r,s=0}^{p-1}, \tag{4.36}$$

where the matrix $\mathbf{H} \in \mathbb{C}^{p \times p}$ is independent of A and B. Moreover, note that

$$c^S - c^B = \begin{cases} -\frac{\text{diam } S}{2} & S \text{ is the left son of } B, \\ \frac{\text{diam } S}{2} & S \text{ is the right son of } B, \end{cases}$$

and we set $\mathbf{C}^{AS} := (\mathbf{D}_1^{AB})^{-1} \mathbf{E}_1^{AS} = \text{diag}\left(e^{\mp \pi i \left(\frac{c^A}{\text{diam } A} + \frac{t_r}{2}\right)/2}\right)_{r=0}^{p-1}$. Altogether, this gives

$$\hat{\mathbf{f}}^{AB} = (\mathbf{D}_2^{AB})^{-1} \mathbf{G}^{-1} \sum_{S \in S_B} \mathbf{C}^{AS} \mathbf{H} \mathbf{E}_2^{AS} \hat{\mathbf{f}}^{PS}, \tag{4.37}$$

where the minus sign is used for the left son and the plus sign for the right son.

4.1.3.2 Lagrange type basis

Our new approach relies more directly on the representation in Equation (4.14) of the interpolating function $\mathcal{J}_p^{AB} g$ for a function $g \in E_B(A)$. In contrast to the variant above using the monomial representation, we apply Equation (4.14) in Equation (4.31). Hence, the functions $f^{PS} \in E_S(P)$, $S \in S_B$, are given by their function values $f^{PS}(x_r^P)$, $r = 0, \ldots, p-1$, at the Chebyshev nodes in P and we compute the function values $f^{AB}(x_r^A)$, $r = 0, \ldots, p-1$, at Chebyshev nodes in A. More detailed, we have

$$f^{AB}(x_r^A) = \sum_{S \in S_B} e^{2\pi i \left(c^S + \frac{\text{diam } S}{2}\right) x_r^A / N} \sum_{s=0}^{p-1} f^{PS}(x_s^P) e^{-2\pi i \left(c^S + \frac{\text{diam } S}{2}\right) x_s^P / N} \cdot l_s^{PS}(x_r^A). \tag{4.38}$$

The Lagrange polynomials are given via the mapping (4.6) by

$$l_s^{PS}(x_r^A) = \prod_{\substack{j=0 \\ j \neq s}}^{p-1} \frac{z(x_r^A) - z(x_j^P)}{z(x_s^P) - z(x_j^P)}. \tag{4.39}$$

Inserting

$$x_r^A = c^A + \frac{t_r}{2} \, \text{diam } A \quad \text{and} \quad x_j^P = c^P + \frac{t_j}{2} \, \text{diam } P$$

and using $\operatorname{diam} P \operatorname{diam} S = N$ in Equation (4.39) yields the representation

$$l_s^{PS}(x_r^A) = \prod_{\substack{j=0 \\ j \neq s}}^{p-1} \frac{e^{-2\pi i \left(c^A - c^P + \frac{t_r}{2}\operatorname{diam} A\right)\frac{\operatorname{diam} S}{N(p-1)}} - e^{-\pi i \frac{t_j}{p-1}}}{e^{-\pi i \frac{t_s}{p-1}} - e^{-\pi i \frac{t_j}{p-1}}}.$$

Furthermore, we have

$$c^A - c^P = \begin{cases} -\frac{\operatorname{diam} A}{2} & A \text{ is the left son of } P, \\ \frac{\operatorname{diam} A}{2} & A \text{ is the right son of } P, \end{cases}$$

and finally

$$l_s^{PS}(x_r^A) = \begin{cases} \displaystyle\prod_{\substack{j=0 \\ j \neq s}}^{p-1} \frac{e^{-\frac{\pi i}{2}\frac{t_r - 1}{p-1}} - e^{-\pi i \frac{t_j}{p-1}}}{e^{-\pi i \frac{t_s}{p-1}} - e^{-\pi i \frac{t_j}{p-1}}} & A \text{ is the left son of } P, \\[2em] \displaystyle\prod_{\substack{j=0 \\ j \neq s}}^{p-1} \frac{e^{-\frac{\pi i}{2}\frac{t_r + 1}{p-1}} - e^{-\pi i \frac{t_j}{p-1}}}{e^{-\pi i \frac{t_s}{p-1}} - e^{-\pi i \frac{t_j}{p-1}}} & A \text{ is the right son of } P. \end{cases} \tag{4.40}$$

With the vectors

$$\mathbf{f}^{PS} = \left(f^{PS}(x_r^P)\right)_{r=0}^{p-1}, \quad \mathbf{f}^{AB} = \left(f^{AB}(x_r^A)\right)_{r=0}^{p-1},$$

the diagonal matrices

$$\mathbf{R}^{AS} := \operatorname{diag}\left(e^{2\pi i \left(c^S + \frac{\operatorname{diam} S}{2}\right)x_r^A/N}\right)_{r=0}^{p-1}, \quad \mathbf{S}^{PS} := \operatorname{diag}\left(e^{-2\pi i \left(c^S + \frac{\operatorname{diam} S}{2}\right)x_s^P/N}\right)_{s=0}^{p-1},$$

and the Lagrange matrix, which depends only on the relation between A and P, whether A is the left or the right son of P,

$$\mathbf{L}^A := \left(l_s^{PS}(x_r^A)\right)_{r,s=0}^{p-1}, \tag{4.41}$$

we finally obtain

$$\mathbf{f}^{AB} = \sum_{S \in S_B} \mathbf{R}^{AS}\mathbf{L}^A\mathbf{S}^{PS}\mathbf{f}^{PS}. \tag{4.42}$$

Note, that we compute the coefficients $\hat{\mathbf{f}}^{AB}$ of the monomial type basis functions and the function values $\mathbf{f}^{AB}$ for the Lagrange type basis.

4.1.4 Stability

While implementing the original scheme [58], we found that the final accuracy of the BSFFT is limited far above machine accuracy as numerically shown in Section 4.5. Of course, the error of the local approximation and thus of the butterfly scheme decreases rapidly with increasing local expansion degree p - at least in precise arithmetic. On the other hand, we show subsequently that the condition number of the interpolation matrix $\mathbf{M}^{AB}$ strongly increases and thus rounding errors take over for larger p. In the following, we prove a weaker increase of the condition number of the Lagrange matrix $\mathbf{L}^A$ which seems to suffice for a stable BSFFT. Based on Lemmata 4.3 and 2.14, we are prepared to prove the following bound on the stability of the local approximation scheme when the monomial type basis is used.

Theorem 4.15. *Let $p \in \mathbb{N}$, $p \geq 3$, and boxes $A := X_{\ell,m}$ and $B := \Omega_{L-\ell,n}$ be given as in Section 4.1.3. The spectral condition number of the interpolation matrix $\mathbf{M}^{AB} \in \mathbb{C}^{p \times p}$ given in Equation (4.33) fulfills*

$$\kappa(\mathbf{M}^{AB}) \geq \begin{cases} \sqrt{p}\left(\dfrac{p-1}{2\pi}\right)^{p-1} & p \geq 3, \hfill (4.43) \\[2ex] \dfrac{1}{\sqrt{p}}\left(\dfrac{2(p-1)}{\pi}\right)^{p-1} & p \geq 3 \text{ and odd.} \hfill (4.44) \end{cases}$$

Proof. We use the factorizations $\mathbf{M}^{AB} = \mathbf{D}_1^{AB}\mathbf{G}\mathbf{D}_2^{AB}$, see Equation (4.34), and

$$\mathbf{G} = \mathbf{D}\mathbf{V}, \quad \mathbf{D} = \mathrm{diag}((\mathrm{e}^{\frac{\pi \mathrm{i}}{2}t_r})_{r=0}^{p-1}), \quad \mathbf{V} = (z_r^s)_{r,s=0}^{p-1}, \quad z_r := \mathrm{e}^{-\pi \mathrm{i}\frac{t_r}{p-1}}.$$

Noting that the norm of all the diagonal matrices and their inverses is equal to one, it suffices to analyze the Vandermonde matrix $\mathbf{V}$. We have

$$\|\mathbf{V}\|_2 = \sup_{\mathbf{x} \in \mathbb{C}^p \setminus \{0\}} \frac{\|\mathbf{V}\mathbf{x}\|_2}{\|\mathbf{x}\|_2} \geq \frac{\|\mathbf{V}\mathbf{e}_0\|_2}{\|\mathbf{e}_0\|_2} = \sqrt{p}$$

for the zeroth unit vector $\mathbf{e}_0 = (1,0,\ldots,0)^\top \in \mathbb{C}^p$ and bound the norm $\|\mathbf{V}^{-1}\|_2$ by a similar technique. Solving the linear system $\mathbf{V}\mathbf{f} = \mathbf{e}_0$ is equivalent to the polynomial interpolation problem

$$q : \mathbb{C} \to \mathbb{C}, \quad q(z) = \sum_{s=0}^{p-1} f_s z^s, \quad \text{such that} \quad q(z_r) = \delta_{r,0} \text{ for } r = 0,\ldots,p-1. \quad (4.45)$$

In terms of the Lagrange polynomials to the nodes $\{z_r\}_{r=0}^{p-1}$, its solution is given by

$$q(z) = \widetilde{L}_0(z) = \prod_{r=1}^{p-1} \frac{z - z_r}{z_0 - z_r} = \left(\prod_{r=1}^{p-1} \frac{1}{z_0 - z_r}\right) z^{p-1} + \ldots$$

and we consider the leading coefficient $f_{p-1} = \prod_{r=1}^{p-1} \frac{1}{z_0 - z_r}$ in the monomial expansion (4.45). Applying Lemma 4.3, estimates (4.9) and (4.10), yields

$$|z_0 - z_r|^2 = 2 - 2\cos\left(\frac{\pi}{p-1}\left(\cos\frac{\pi}{2p} - \cos\frac{2r+1}{2p}\pi\right)\right)$$

$$\leq 2 - 2\cos\frac{2\pi}{p-1} \leq \left(\frac{2\pi}{p-1}\right)^2.$$

Hence, we obtain

$$\|\mathbf{V}^{-1}\|_2 \geq \|\mathbf{V}^{-1}\mathbf{e}_0\|_2 \geq |f_{p-1}| = \prod_{r=1}^{p-1} \frac{1}{|z_0 - z_r|} \geq \left(\frac{p-1}{2\pi}\right)^{p-1}$$

and thus the assertion (4.43).

For $p = 2n + 1$, $n \in \mathbb{N}$, we consider the linear system $\mathbf{V}f = \mathbf{e}_n$ with the n-th unit vector, which is equivalent to the interpolation problem

$$q : \mathbb{C} \to \mathbb{C}, \quad q(z) = \sum_{s=0}^{p-1} f_s z^s, \quad \text{such that} \quad q(z_r) = \delta_{n,r} \text{ for } r = 0, \dots, p-1.$$

Noting $z_n = 1$ and analogously to the above consideration, we have

$$q(z) = \widetilde{L}_n(z) = \prod_{\substack{k=0 \\ k \neq n}}^{p-1} \frac{z - z_k}{1 - z_k} = f_{p-1} z^{p-1} + \dots \quad , \quad f_{p-1} = \prod_{\substack{k=0 \\ k \neq n}}^{p-1} \frac{1}{1 - z_k}.$$

Using Lemma 4.3 (4.10) yields

$$|1 - z_k|^2 = 2 - 2\cos\frac{\pi}{p-1} t_k \leq \frac{\pi^2}{(p-1)^2} t_k^2$$

and together with Lemma 2.10 (2.9) in

$$\|\mathbf{V}^{-1}\|_2 \geq |f_{p-1}| \geq \left(\frac{p-1}{\pi}\right)^{p-1} \prod_{\substack{k=0 \\ k \neq n}}^{p-1} |t_k|^{-1} = \frac{2^{p-1}}{p} \left(\frac{p-1}{\pi}\right)^{p-1}$$

the assertion (4.44). □

The condition number of the matrix $\mathbf{N}^{AS}$ in Section 4.1.3.1 can be analyzed in the same way to yield $\kappa(\mathbf{N}^{AS}) \approx 2^{p-1} \kappa(\mathbf{M}^{AB})$. In contrast to the lower bound on the conditioning of the original method, we obtain an upper bound for the local approximation scheme when the Lagrange type basis, cf. Section 4.1.3.2, is used.

Theorem 4.16. *Under the assumptions of Theorem 4.15, the spectral condition number of the Lagrange interpolation matrix $\mathbf{L}^A \in \mathbb{C}^{p \times p}$ given in Equation (4.41) fulfills*

$$\kappa(\mathbf{L}^A) \leq K_p^2 \frac{\sqrt{2p \cdot 34^p}}{4} \leq \frac{\sqrt{2p}}{4} \cdot 6^{p+1}. \tag{4.46}$$

Proof. The Lagrange functions (4.40) are independent of the box A, up to its relation to the father box P. Setting $x_r = \frac{t_r}{2} \mp \frac{1}{2}$ in Equation (4.40) yields

$$|l_s^{PS}(x_r^A)|^2 = \prod_{\substack{j=0 \\ j \neq s}}^{p-1} \frac{1 - \cos\left(\frac{\pi}{p-1}(t_j - x_r)\right)}{1 - \cos\left(\frac{\pi}{p-1}(t_j - t_s)\right)}.$$

Since $x_r \in [-1, 1]$ and thus $t_j - x_r \in [-2, 2]$, we can follow the ideas in the proof of Lemma 4.4 to obtain the relation

$$\leq K_p^2 \prod_{\substack{j=0 \\ j \neq s}}^{p-1} \left(\frac{x_r - t_j}{t_s - t_j}\right)^2 = K_p^2 \cdot (L_s(x_r))^2.$$

Due to Inequality (2.12) in Lemma 2.14, this yields

$$\|\mathbf{L}^A\|_{\mathrm{F}}^2 = \sum_{r=0}^{p-1}\sum_{s=0}^{p-1}|l_s^{PS}(x_r^A)|^2 \leq K_p^2 \sum_{r=0}^{p-1}\max_{x\in[-1,1]}\sum_{s=0}^{p-1}(L_s(x))^2 \leq 2pK_p^2. \tag{4.47}$$

In view of the related polynomial interpolation problem on the complex unit circle and by changing the basis of the Lagrange polynomials l_s^{PS} to l_s^{AS}, the entries of the inverse Lagrange matrix can be written as

$$\left(\left(\mathbf{L}^A\right)^{-1}\right)_{r,s} = l_s^{AS}(x_r^P) = \begin{cases} \prod_{\substack{j=0 \\ j\neq s}}^{p-1} \dfrac{e^{-\pi i\left(t_r+\frac{1}{2}\right)/(p-1)}-e^{-\pi i\frac{t_j}{2(p-1)}}}{e^{-\pi i\frac{t_s}{2(p-1)}}-e^{-\pi i\frac{t_j}{2(p-1)}}} & A \text{ is the left son of } P, \\ \prod_{\substack{j=0 \\ j\neq s}}^{p-1} \dfrac{e^{-\pi i\left(t_r-\frac{1}{2}\right)/(p-1)}-e^{-\pi i\frac{t_j}{2(p-1)}}}{e^{-\pi i\frac{t_s}{2(p-1)}}-e^{-\pi i\frac{t_j}{2(p-1)}}} & A \text{ is the right son of } P. \end{cases}$$

Analogously to the first part of the proof, we set $x_r = t_r \pm \frac{1}{2} \subset \left[-\frac{3}{2},\frac{3}{2}\right]$ and use $1/4t_j - x_r/2 \in [-1,1]$ to use again the ideas in proof of Lemma 4.4. Together with $2x_r \in [-3,3]$, Inequality (2.14) in Lemma 2.14 yields

$$|l_s^{AS}(x_r^P)|^2 = \prod_{\substack{j=0 \\ j\neq s}}^{p-1} \frac{1 - \cos\left(\frac{2\pi}{p-1}\left(\frac{t_j}{4}-\frac{x_r}{2}\right)\right)}{1 - \cos\left(\frac{2\pi}{p-1}\left(\frac{t_j}{4}-\frac{t_s}{4}\right)\right)} \leq K_p^2 \cdot (L_s(2x_r))^2 \leq \frac{34^p K_p^2}{16p^2}.$$

Combining Equation (4.47) and the slightly simpler estimate

$$\left\|\left(\mathbf{L}^A\right)^{-1}\right\|_{\mathrm{F}}^2 = \sum_{r=0}^{p-1}\sum_{s=0}^{p-1}|l_s^{AS}(x_r^P)|^2 \leq \frac{34^p K_p^2}{16},$$

the assertion follows by bounding the spectral norm by the Frobenius norm. Setting $x = \frac{2\pi}{p-1}$ in Lemma 4.3 (4.12) finally yields $K_p^2\sqrt{34} \leq 6^2$. □

4.2 The multivariate case

We develop the method of the multi-dimensional butterfly fast Fourier transform from the one-dimensional case. We will define the interpolation in the multivariate case via tensor products and consider the realizations in the monomial and Lagrange type basis.

4.2.1 Interpolation via tensor products

Let the spatial dimension $d \in \mathbb{N}$, $d \geq 2$, be given. Moreover, let the spatial domain X and the frequency domain Ω be given with some node sets $\widetilde{X}$ and $\widetilde{\Omega}$ as introduced at the beginning of this chapter. Let two intervals $D_1, D_2 \subset \mathbb{R}$ be given. We define the tensor product of two functions $u \in C(D_1)$, $v \in C(D_2)$ by

$$(u \otimes v)(x,y) := u(x)v(y).$$

For d-dimensional boxes $A = A_1 \times A_2 \times \cdots \times A_d$, $B = B_1 \times B_2 \times \cdots \times B_d$, and $p \in \mathbb{N}$, $p \geq 2$, we introduce an equispaced multi grid by

$$\boldsymbol{\xi}_{\mathbf{s}}^B := \left(\xi_{s_1}^{B_1}, \ldots, \xi_{s_d}^{B_d}\right)^\top, \quad s_1, \ldots, s_d \in \{0, \ldots, p-1\}, \tag{4.48}$$

where $\xi_{s_\nu}^{B_\nu}$, $s_\nu = 0, \ldots, p-1$, are the equispaced nodes in the box B_ν. Moreover, we define the tensor product spaces

$$
\begin{aligned}
E_B(A) &:= E_{B_1}(A_1) \otimes \cdots \otimes E_{B_d}(A_d) \\
&= \left\{ h : \exists m \in \mathbb{N}, u_i^1 \in E_{B_1}(A_1), \ldots, u_i^d \in E_{B_d}(A_d) \text{ with } h = \sum_{i=0}^m u_i^1 \otimes \cdots \otimes u_i^d \right\} \\
&= \left\{ h : A \to \mathbb{C} : h(\boldsymbol{x}) = \sum_{k=0}^n \hat{h}_k e^{2\pi i \boldsymbol{\xi}_k \boldsymbol{x}/N}, \ n \in \mathbb{N}, \ \hat{h}_k \in \mathbb{C}, \ \boldsymbol{\xi}_k \in B \right\}, \\
E_B^p(A) &:= E_{B_1}^p(A_1) \otimes \cdots \otimes E_{B_d}^p(A_d) \\
&= \left\{ h : \exists m \in \mathbb{N}, u_i^1 \in E_{B_1}^p(A_1), \ldots, u_i^d \in E_{B_d}^p(A_d) \text{ with } h = \sum_{i=0}^m u_i^1 \otimes \cdots \otimes u_i^d \right\} \\
&= \left\{ h : A \to \mathbb{C} : h(\boldsymbol{x}) = \sum_{s_1,\ldots,s_d=0}^{p-1} \hat{h}_{s_1,\ldots,s_d} e^{2\pi i \boldsymbol{\xi}_{\mathbf{s}}^B \boldsymbol{x}/N}, \ \hat{h}_{s_1,\ldots,s_d} \in \mathbb{C}, \right\}.
\end{aligned}
$$

We extend the definition of the trigonometric interpolation, see Definition 4.1 to higher dimensions.

Definition 4.17 (Multi-dimensional trigonometric interpolation). *Let $p \in \mathbb{N}$, $p \geq 2$, be given. Furthermore, let $A \subset X$ and $B \subset \Omega$ two d-dimensional boxes.*

(i) The Chebyshev tensor grid in A is defined by

$$\boldsymbol{x}_{\mathbf{r}}^A := (x_{r_1}^{A_1}, \ldots, x_{r_d}^{A_d})^\top, \quad r_1, \ldots, r_d \in \{0, \ldots, p-1\},$$

where $x_{r_\nu}^{A_\nu}$ are the Chebyshev nodes in the box A_ν.

(ii) To shorten notation we define the shift function $d^B : A \to \mathbb{C}$ by

$$d^B(\boldsymbol{x}) := e^{2\pi i \left(\mathbf{c}^B + \frac{\mathbf{w}^B}{2}\right)\boldsymbol{x}/N},$$

where $\mathbf{w}^B := (\operatorname{diam} B_1, \ldots, \operatorname{diam} B_d)^\top$. Moreover, the interpolation operator $\mathcal{J}_p^{AB} : C(A) \to E_B^p(A)$ for complex valued functions is defined as tensor product of the one-dimensional interpolation operators by

$$
\begin{aligned}
\left(\mathcal{J}_p^{AB} g\right)(\boldsymbol{x}) &:= \left(\bigotimes_{\nu=1}^d \mathcal{J}_p^{A_\nu B_\nu} g\right)(\boldsymbol{x}) \tag{4.49} \\
&:= d^B(\boldsymbol{x}) \sum_{r_1,\ldots,r_2=0}^{p-1} g(\boldsymbol{x}_{\mathbf{r}}^A) d^B(-\boldsymbol{x}_{\mathbf{r}}^A) \prod_{\nu=1}^d l_{r_\nu}^{A_\nu B_\nu}(x_\nu),
\end{aligned}
$$

where $\boldsymbol{x} = (x_1, \ldots, x_d)^\top$.

(iii) We set $\boldsymbol{x}_{1,r} := (x_r^{A_1}, x_2, \ldots, x_d)$, $\bar{A} := A_2 \times \cdots \times A_d$, $\bar{B} := B_2 \times \cdots \times B_d$, $\boldsymbol{y} := (x_2, \ldots, x_d)$ and $\boldsymbol{y}_{\bar{\mathbf{r}}} := (x_{r_2}^{A_2}, \ldots, x_{r_d}^{A_d})$. We denote the identity in one direction by $\mathcal{I}$ and $\mathcal{I}_\nu := \overset{\nu}{\underset{k=1}{\bigotimes}} \mathcal{I}$. Moreover, we define the trigonometric interpolation in one direction

$$\left(\mathcal{J}_p^{A_1 B_1} \otimes \mathcal{I}_{d-1}\right) g(\boldsymbol{x}) := d^{B_1}(x_1) \sum_{r=0}^p g(\boldsymbol{x}_{1,r}) d^{B_1}(-x_r^{A_1}) l_r^{A_1 B_1}(x_1)$$

and the trigonometric interpolation in the last $d-1$ directions by

$$\left(\mathcal{I} \otimes \mathcal{J}_p^{\bar{A}\bar{B}}\right) g(\boldsymbol{x}) := d^{\bar{B}}(\boldsymbol{y}) \sum_{r_2,\ldots,r_d=0}^p g(x_1, \boldsymbol{y}_{\bar{\mathbf{r}}}) d^{\bar{B}}(\boldsymbol{y}_{\bar{\mathbf{r}}}) \prod_{\nu=2}^d l_{r_\nu}^{A_\nu B_\nu}(x_\nu).$$

The interpolation can be written as the composition of the interpolation in the first coordinate and the interpolation in the last $d-1$ coordinates. The operator norm of the interpolation in the first coordinate is bounded by the Lebesgue constant which increases logarithmically for increasing approximation rank.

Lemma 4.18. *Let $p \in \mathbb{N}$, $p \geq 3$, and two d-dimensional boxes $A \subset X$, $B \subset \Omega$ be given. For $g : A \to \mathbb{C}$, $g \in C(A)$, it follows*

$$\left(\mathcal{J}_p^{A_1 B_1} \otimes \mathcal{I}_{d-1}\right) \left(\mathcal{I} \otimes \mathcal{J}_p^{\bar{A}\bar{B}}\right) g = \mathcal{J}_p^{AB} g.$$

and

$$\left\|\mathcal{J}_p^{A_1 B_1} \otimes \mathcal{I}_{d-1}\right\| := \max_{g \in C(A)\backslash\{0\}} \frac{\left\|(\mathcal{J}_p^{A_1 B_1} \otimes \mathcal{I}_{d-1})g\right\|_{C(A)}}{\|g\|_{C(A)}} \leq C_p$$

with C_p given by Equation (4.15).

Proof. Applying the definitions yields the first assertion

$$\left(\mathcal{J}_p^{A_1 B_1} \otimes \mathcal{I}_{d-1}\right) \left(d^{\bar{B}}(\boldsymbol{y}) \sum_{r_2,\ldots,r_d=0}^p g(x_1, \boldsymbol{y}_{\bar{\mathbf{r}}}) d^{\bar{B}}(-\boldsymbol{y}_{\bar{\mathbf{r}}}) \prod_{\nu=2}^d l_{r_\nu}^{A_\nu B_\nu}(x_\nu)\right)$$

$$= d^{\bar{B}}(\boldsymbol{y}) \sum_{r_2,\ldots,r_d=0}^p \left(\left(\mathcal{J}_p^{A_1 B_1} \otimes \mathcal{I}_{d-1}\right) g\right)(x_1, \boldsymbol{y}_{\bar{\mathbf{r}}}) d^{\bar{B}}(-\boldsymbol{y}_{\bar{\mathbf{r}}}) \prod_{\nu=2}^d l_{r_\nu}^{A_\nu B_\nu}(x_\nu)$$

$$= d^{\bar{B}}(\boldsymbol{y}) \sum_{r_2,\ldots,r_d=0}^p \left(d^{B_1}(x_1) \sum_{r_1=0}^p g(\boldsymbol{x}_{\mathbf{r}}^A) d^{B_1}(-x_{r_1}^{A_1}) l_{r_1}^{A_1 B_1}(x_1)\right) d^{\bar{B}}(-\boldsymbol{y}_{\bar{\mathbf{r}}}) \prod_{\nu=2}^d l_{r_\nu}^{A_\nu B_\nu}(x_\nu)$$

$$= \mathcal{J}_p^{AB} g(\boldsymbol{x}).$$

Let now $g \in E_B(A)$, $g(\boldsymbol{x}) = \sum_{k=0}^n \hat{g}_k e^{2\pi i \boldsymbol{\xi}_k \boldsymbol{x}/N}$, where $\boldsymbol{\xi}_k \in B$ and $n \in \mathbb{N}$. We set $\boldsymbol{x} := (x, \boldsymbol{y})^\top$ and $v_{\boldsymbol{y}} \colon A_1 \to \mathbb{C}$, $v_{\boldsymbol{y}}(x) := g(x, \boldsymbol{y})$. The function $(\mathcal{J}_p^{A_1 B_1} \otimes \mathcal{I}_d)g$ is continuous and has an absolute maximum in A. We choose $\boldsymbol{x}^* := (x^*, \boldsymbol{y}^*)^\top \in A$ such that $(\mathcal{J}_p^{A_1 B_1} \otimes \mathcal{I}_d)g$ has its absolute maximum in $\boldsymbol{x}^*$. Hence, it follows

$$\max_{\boldsymbol{x} \in A} \left|(\mathcal{J}_p^{A_1 B_1} \otimes \mathcal{I}_d)g(\boldsymbol{x})\right| = \left|(\mathcal{J}_p^{A_1 B_1} \otimes \mathcal{I}_d)g(\boldsymbol{x}^*)\right| = \max_{x \in A_1} \left|\mathcal{J}_p^{A_1 B_1} v_{\boldsymbol{y}^*}(x)\right|.$$

We set $\boldsymbol{\xi}_k = (\xi_k, \boldsymbol{\eta}_k)^\top$ and obtain

$$v_{\boldsymbol{y}^*}(x) = \sum_{k=0}^{n} f_k \mathrm{e}^{2\pi\mathrm{i}\xi_k x/N} \mathrm{e}^{2\pi\mathrm{i}\boldsymbol{\eta}_k \boldsymbol{y}^*/N}.$$

Since we have $v_{\boldsymbol{y}^*} \in E_{B_1}$, we can apply Equation (4.15) and obtain the assertion

$$\max_{x \in A_1} \left| \mathcal{J}_p^{A_1 B_1} v_{\boldsymbol{y}^*}(x) \right| \leq C_p \max_{x \in A_1} |v_{\boldsymbol{y}^*}(x)| = C_p \max_{\boldsymbol{x} \in A} g(\boldsymbol{x}).$$

$\square$

Theorem 4.19. *Let* $d, p, N \in \mathbb{N}$, $p \geq 5$, *two boxes* $A, B \subset \mathbb{R}^d$ *be admissible in the sense*

$$\operatorname{diam} A \operatorname{diam} B \leq N,$$

and $g \in E_B(A)$, $g(\boldsymbol{x}) := \sum_{k=0}^{n} \hat{g}_k \mathrm{e}^{2\pi\mathrm{i}\boldsymbol{\xi}_k \boldsymbol{x}/N}$, $\hat{g}_k \in \mathbb{C}$, $k = 1, \ldots, M_2$, *then we have the error estimate*

$$\left\| g - \mathcal{J}_p^{AB} g \right\|_{C(A)} \leq \frac{(1+C_p)(C_p^d - 1)}{C_p - 1} \cdot c_p \cdot \|\hat{\boldsymbol{g}}\|_1,$$

where the constants are given by Equation (4.15) and Equation (4.19), respectively.

Proof. The assertion follows from Theorem 4.19 for $d = 1$. We assume, that the assertion holds true for a $d \in \mathbb{N}$. Inductively, we consider two $(d+1)$-dimensional boxes $A = A_1 \times \cdots \times A_d \times A_{d+1}$ and $B = B_1 \times \cdots \times B_d \times B_{d+1}$. We set $\bar{A} = A_2 \times \cdots \times A_{d+1}$ and $\bar{B} = B_2 \times \cdots \times B_{d+1}$. Applying the triangle inequality and extracting the norm of the interpolation operator yields

$$\begin{aligned}
\left\| g - \mathcal{J}_p^{AB} g \right\|_{C(A)} &= \left\| g - \left(\mathcal{J}_p^{A_1 B_1} \otimes \mathcal{I}_d \right) \left(\mathcal{I} \otimes \mathcal{J}_p^{\bar{A}\bar{B}} \right) g \right\|_{C(A)} \\
&\leq \left\| g - \left(\mathcal{J}_p^{A_1 B_1} \otimes \mathcal{I}_d \right) g \right\|_{C(A)} + \\
&\quad + \left\| \mathcal{J}_p^{A_1 B_1} \otimes \mathcal{I}_d \right\| \left\| g - \left(\mathcal{I} \otimes \mathcal{J}_p^{\bar{A}\bar{B}} \right) g \right\|_{C(A)}.
\end{aligned}$$

Choosing the maximal points of $\left\| g - \left(\mathcal{J}_p^{A_1 B_1} \otimes \mathcal{I}_d \right) g \right\|_{C(A)}$ and $\left\| g - \left(\mathcal{I} \otimes \mathcal{J}_p^{\bar{A}\bar{B}} \right) g \right\|_{C(A)}$, respectively, we just have to consider the one- and accordingly the d-dimensional interpolation. By applying Theorem 4.8, it follows

$$\left\| g - \left(\mathcal{J}_p^{A_1 B_1} \otimes \mathcal{I}_d \right) g \right\|_{C(A)} \leq (1 + C_p) c_p \|\hat{\boldsymbol{g}}\|_1$$

and due to the induction hypothesis

$$\left\| g - \left(\mathcal{I} \otimes \mathcal{J}_p^{\bar{A}\bar{B}} \right) g \right\|_{C(A)} \leq \frac{(1+C_p)(C_p^d - 1)}{C_p - 1} \cdot c_p \cdot \|\hat{\boldsymbol{g}}\|_1.$$

Altogether, the assertion follows with Lemma 4.18

$$\left\| g - \mathcal{J}_p^{AB} g \right\|_{C(A)} \leq (1 + C_p) c_p \|\hat{\boldsymbol{g}}\|_1 + C_p \frac{(1+C_p)(C_p^d - 1)}{C_p - 1} \cdot c_p \cdot \|\hat{\boldsymbol{g}}\|_1.$$

$\square$

4.2.2 Butterfly algorithm

We expand Algorithm 3 to higher dimensions. The dyadic decompositions of the domains $X = [0, N]^d$ and $\Omega = [0, N]^d$ are given by

$$\begin{aligned} X_{\ell,\mathbf{m}} &:= X_{\ell,m_1} \times \cdots \times X_{\ell,m_d}, \\ \Omega_{\ell,\mathbf{n}} &:= \Omega_{\ell,n_1} \times \cdots \times \Omega_{\ell,n_d}, \end{aligned} \tag{4.50}$$

where $\mathbf{m} := (m_1, \ldots, m_d) \in \mathbb{N}_0^d$ with $\|\mathbf{m}\|_\infty < 2^\ell$ and $\mathbf{n} := (n_1, \ldots, n_d) \in \mathbb{N}_0^d$ with $\|\mathbf{n}\|_\infty < 2^\ell$ are the index vectors, which declare the position of the boxes $X_{\ell,\mathbf{m}}$ and $\Omega_{\ell,\mathbf{n}}$, respectively. Each level ℓ has N^d admissible pairs $(X_{\ell,\mathbf{m}}, \Omega_{L-\ell,\mathbf{n}})$. Similar to the one-dimensional case, the result of the dyadic decomposition of X and Ω is the formation of the trees $\mathcal{T}_X$ and $\mathcal{T}_\Omega$, where the root nodes are X and Ω. If the box $X_{\ell,\mathbf{m}}$ with $\|\mathbf{m}\|_\infty < 2^\ell$ is a node in the tree $\mathcal{T}_X$, then the child box $X_{\ell+1,2\mathbf{m}+\mathbf{r}}$, $\mathbf{r} \in \{0,1\}^d$, is a node in the tree $\mathcal{T}_X$ if and only if $X_{\ell+1,2\mathbf{m}+\mathbf{r}} \cap \widetilde{X} \neq \emptyset$, and it is connected consequently by an edge to the parent node $X_{\ell,\mathbf{m}}$. Each node of the tree in level $\ell = 1, \ldots, L-1$ has at most 2^d sons. The tree $\mathcal{T}_\Omega$ is built up analogously. The product of the trees $\mathcal{T}_X$ and $\mathcal{T}_\Omega$ is a butterfly graph, which is built up as in the one-dimensional case, see the previous Section 4.1.2 for more details. The complete butterfly graph has $L + 1$ levels and N^d nodes in each level. It must be pointed out that the number of nodes in the trees and accordingly in the butterfly graph is dependent on the position of the sampling nodes $\widetilde{X}$ and $\widetilde{\Omega}$. We will see later in Section 4.4, that the computational and space complexity depends on the position of the sampling nodes. To be more precise the complexity of the algorithm depends on the number of nonempty boxes. We write $A \in \mathcal{T}_X$, if A is a node in the tree $\mathcal{T}_X$, analogously for $B \in \mathcal{T}_\Omega$. The trees are realized as a recursive data structure, see Algorithm 4, which is a generalization of Algorithm 2. The tree $\mathcal{T}_X$ can be generated via the constructor

$$\mathcal{T}_X := \text{TREE}(\widetilde{X}, (1, \ldots, M_1), 0, \mathbf{0}).$$

We denote the floor function for a vector $\mathbf{v} \in \mathbb{R}^d$ by $\lfloor \mathbf{v} \rfloor := (\lfloor v_1 \rfloor, \ldots, \lfloor v_d \rfloor)$. Algorithm 5 is a generalization of Algorithm 3 to higher dimensions. For simplicity of notation, Algorithm 5 traverses the full butterfly graph. We describe in Section 4.4 the details of the conditions on the nodes $\widetilde{X}$ and $\widetilde{\Omega}$, that we call the algorithm sparse. We denote Algorithm 5 short as BSFFT.

4.2.3 Realization

According to the realization in the one-dimensional case, see Section 4.1.3, we consider the realization in the d-dimensional case in Algorithm 5 Step 3. The boxes in X and Ω have the relations $X_{\ell,\mathbf{m}} \subset X_{\ell-1,\lfloor \mathbf{m}/2 \rfloor}$ and $\Omega_{L-\ell,\mathbf{n}} = \cup_{\mathbf{k} \in \{0,1\}^d} = \Omega_{L-\ell,2\mathbf{n}+\mathbf{k}}$. Again, to shorten notation let $A := A_1 \times \cdots \times A_d = X_{\ell,\mathbf{m}}$ with $A_i := X_{\ell,m_i}$ and $B := B_1 \times \cdots \times B_d = \Omega_{L-\ell,\mathbf{n}}$ with $B_i := \Omega_{L-\ell,n_i}$ for $i = 1, \ldots, d$ and for $\ell \in \{1, \ldots, L\}$, $\mathbf{m} \in \mathbb{N}_0^d$ with $\|\mathbf{m}\|_\infty < 2^\ell$ and $\mathbf{n} \in \mathbb{N}_0^d$ with $\|\mathbf{n}\|_\infty < 2^{L-\ell}$. In Step 3, we have to compute the approximation $f^{AB} \in E_B(A)$ of (4.31), where now S_{B_i} denotes the set of sons of the box B_i for $i = 1, \ldots, d$ and

$$S_B := \{S = S_1 \times \cdots \times S_d : (S_1 \times \cdots \times S_d) \cap \widetilde{\Omega} \neq \emptyset, \ S_1 \in S_{B_1}, \ldots, S_d \in S_{B_d}\}$$

Algorithm 4 Constructor tree

Input:

$d \in \mathbb{N}$	▷ dimension
$N = 2^L$, $L \in \mathbb{N}$	▷ bandwidth
$M_1 \in \mathbb{N}$	▷ number of sampling nodes
$\widetilde{X} = \{\mathbf{x}_1, \ldots, \mathbf{x}_{M_1}\} \subset X$	▷ sampling nodes
$I \subset \{1, \ldots, M_1\}$	▷ index set
$\ell \in \{0, 1, \ldots, L\}$	▷ level
$\mathbf{m} \in \{0, 1, \ldots, 2^\ell - 1\}^d$	▷ position parameter

> **function** TREE($\widetilde{X}, I, N, \ell, \mathbf{m}$)
>> $\mathcal{T}.\ell := \ell$
>> $\mathcal{T}.\mathbf{m} := \mathbf{m}$
>> **if** $2^{\mathcal{T}.\ell} = N$ or $I = \emptyset$ **then**
>>> $\mathcal{T}.I := I$
>> **else**
>>> **for** $\mathbf{r} \in \{0, 1\}^d$ **do**
>>>> $I_\mathbf{r} := \{j \in I : \mathbf{x}_j \in X_{\ell+1, 2\mathbf{m}+\mathbf{r}}\}$
>>>> **if** $I_{r_1, \ldots, r_d} \neq \emptyset$ **then**
>>>>> $s := \text{TREE}(\widetilde{X}, I_\mathbf{r}, \ell+1, 2\mathbf{m}+\mathbf{r})$ ▷ son of data type TREE
>>>>> $\mathcal{T}.S := \{\mathcal{T}.S, s\}$ ▷ set of sons
>>>> **end if**
>>> **end for**
>> **end if**
> **end function**

Output:

$\mathcal{T}$	▷ tree

Algorithm 5 Butterfly sparse fast Fourier transform in higher dimensions (BSFFT)

Input:

$d, L, M_1, M_2, p \in \mathbb{N}, \, p \geq 2, \, N = 2^L,$
$\hat{f}_k \in \mathbb{C}, \, k = 1, \ldots, M_2,$
$\Omega = \{\boldsymbol{\xi}_k \in [0, N]^d : k = 1, \ldots, M_2\},$
$\widetilde{X} = \{\boldsymbol{x}_j \in [0, N]^d : j = 1, \ldots, M_1\}.$

---

1. Decompose the space and frequency domain X and Ω dyadically, see Equation (4.50).

2. Define the functions $f^{\Omega_{L,\mathbf{n}}} : X \rightarrow \mathbb{C}$ and the corresponding interpolating functions $f^{X_{0,0}\Omega_{L,\mathbf{n}}}$ by

   for $n_1 \ldots, n_d = 0, \ldots, 2^L - 1$ **do**

   $$f^{\Omega_{L,\mathbf{n}}}(\boldsymbol{x}) := \sum_{\boldsymbol{\xi}_j \in \Omega_{L,\mathbf{n}} \cap \widetilde{\Omega}} \hat{f}_j \mathrm{e}^{2\pi \mathrm{i} \boldsymbol{\xi}_j \boldsymbol{x} / N}, \tag{4.51}$$
   $$f^{X_{0,0}\Omega_{L,\mathbf{n}}} := \mathcal{J}_p^{X_{0,0}\Omega_{L,\mathbf{n}}} f^{\Omega_{L,\mathbf{n}}}.$$

   end for

3. Compute for the levels $\ell = 1, \ldots, L$ the recursively defined interpolating functions $f^{X_{\ell,\mathbf{m}}\Omega_{L-\ell,\mathbf{n}}} : X_{\ell,\mathbf{m}} \rightarrow \mathbb{C}.$

   for $\ell = 1, \ldots, L$ **do**
       for $m_1, \ldots, m_d = 0, \ldots, 2^\ell - 1$ and $n_1, \ldots, n_d = 0, \ldots, 2^{L-\ell} - 1$ **do**

   $$f^{X_{\ell,\mathbf{m}}\Omega_{L-\ell,\mathbf{n}}} := \mathcal{J}_p^{X_{\ell,\mathbf{m}}\Omega_{L-\ell,\mathbf{n}}} \sum_{\mathbf{r} \in \{0,1\}^d} f^{X_{\ell-1,\lfloor \mathbf{m}/2 \rfloor}\Omega_{L-\ell+1,2\mathbf{n}+\mathbf{r}}}, \tag{4.52}$$

       end for
   end for

4. The final approximation $\widetilde{f}$ of $f \in E_\Omega(X)$ is a composed function of approximating functions from the L-th level and given by

   for $j = 1, \ldots, M_1$ **do**

   $$\mathbf{m} := \lfloor \boldsymbol{x}_j \rfloor$$
   $$\widetilde{f}(\boldsymbol{x}_j) := f^{X_{L,\mathbf{m}}\Omega_{0,0}}(\boldsymbol{x}_j). \tag{4.53}$$

   end for

Output:
  Approximate function values $\widetilde{f}(\boldsymbol{x}_j), \, j = 1, \ldots, M_1.$

---

denotes the set of sons, at most 2^d, of the d-dimensional box B and the interpolation error is small in each of the spatial sets A, again at most 2^d. We denote the father box of A by P again. We rely on the same admissible condition as in the one-dimensional case, diam A diam $B = N$, and on the result of the dyadic decomposition which results in diam P diam $B = 2N$. In this section, we present the calculation rules for computing the function values of f^{AB}. We consider the complexity of these formulas in Section 4.4.

4.2.3.1 Monomial type basis

Let $g \in E_B(A)$. The interpolation in Definition 4.17 has a unique solution in $E_B^p(A)$. The interpolation representations (4.54) and (4.49) are equivalent.

Theorem 4.20 (Monomial representation.)**.** *Let $p \in \mathbb{N}$, $p \geq 2$, and two d-dimensional boxes $A \subset X$, $B \subset \Omega$ be given. The interpolation $\mathcal{J}_p^{AB} g$ for a function $g \in C(A)$ has the representation*

$$\mathcal{J}_p^{AB} g(\boldsymbol{x}) = \sum_{s_1,\ldots,s_d=0}^{p-1} \hat{g}_{s_1,\ldots,s_d}^{AB} e^{2\pi i \boldsymbol{\xi}_{\mathbf{s}}^B \boldsymbol{x}/N}, \tag{4.54}$$

with $\boldsymbol{\xi}_{\mathbf{s}}^B$ given by Equation (4.48) and with coefficients $\hat{g}_{s_1,\ldots,s_d}^{AB}$ given by

$$\hat{\mathbf{g}}^{AB} := \left(\hat{g}_{s_1,\ldots,s_d}^{AB}\right)_{s_1,\ldots,s_d=0,\ldots,p-1} = (\mathbf{M}^{AB})^{-1} \mathbf{g}^{AB} \in \mathbb{C}^{p^d} \tag{4.55}$$

with the interpolation matrix

$$\mathbf{M}^{AB} := \bigotimes_{\nu=1}^{d} \mathbf{M}^{A_\nu B_\nu} = \left(e^{2\pi i \boldsymbol{\xi}_{\mathbf{s}}^B \boldsymbol{x}_{\mathbf{r}}^A/N}\right)_{r_1,\ldots,r_d;s_1,\ldots,s_d=0}^{p-1} \in \mathbb{C}^{p^d \times p^d}, \; and$$

$$\mathbf{g}^{AB} := \left(g(\boldsymbol{x}_{\mathbf{r}}^A)\right)_{r_1,\ldots,r_d=0,\ldots,p-1} \in \mathbb{C}^{p^d}.$$

Proof. The assertion follows from Lemma 4.4 for $d = 1$. We assume that the assertion holds true for $d \in \mathbb{N}$. Inductively, let $A = A_1 \times \cdots \times A_{d+1}$, $B = B_1 \times \cdots \times B_{d+1}$, $\bar{A} = A_2 \times \cdots \times A_{d+1}$, $\bar{B} = B_2 \times \cdots \times B_{d+1}$, and $\boldsymbol{y} = (x_2,\ldots,x_{d+1})$. Let $g \in C(A)$, we define for $x_1 \in A_1$, $v_{x_1} \colon \bar{A} \to \mathbb{C}$, $v_{x_1}(\boldsymbol{y}) = g(x_1, \boldsymbol{y})$. For x_1 fixed, we have

$$\mathcal{J}_p^{AB} g(\boldsymbol{x}) = \left(\mathcal{J}_p^{A_1 B_1} \otimes \mathcal{I}_d\right) \left(\mathcal{I} \otimes \mathcal{J}_p^{\bar{A}\bar{B}}\right) g(x_1, \boldsymbol{y}) = \left(\mathcal{J}_p^{A_1 B_1} \otimes \mathcal{I}_d\right) \mathcal{J}_p^{\bar{A}\bar{B}} v_{x_1}(\boldsymbol{y}).$$

We set the index vectors $\bar{\mathbf{s}} = (s_2,\ldots,s_{d+1})$, $\bar{\mathbf{r}} = (r_2,\ldots,r_{d+1})$, and we declare $\boldsymbol{y}_{\bar{\mathbf{r}}}^{\bar{A}}$ as the Chebyshev grid in $\bar{A}$. Due to the induction hypothesis, the matrix $\mathbf{M}^{\bar{A}\bar{B}}$ is regular. We define the vector of coefficients by

$$\hat{\mathbf{g}}^{\bar{A}\bar{B}}(x_1) := \left(\hat{g}_{\bar{\mathbf{s}}}^{\bar{A}\bar{B}}(x_1)\right)_{s_2,\ldots,s_{d+1}=0,\ldots,p-1} := (\mathbf{M}^{\bar{A}\bar{B}})^{-1} \boldsymbol{v}_{x_1}^{\bar{A}\bar{B}},$$

where $\boldsymbol{v}_{x_1}^{\bar{A}\bar{B}} := \left(v_{x_1}(\boldsymbol{y}_{\bar{\mathbf{r}}}^{\bar{A}})\right)_{r_2,\ldots,r_{d+1}=0,\ldots,p-1}$. Applying the induction hypothesis yields

$$\mathcal{J}_p^{\bar{A}\bar{B}} v_{x_1}(\boldsymbol{y}) = \sum_{s_2,\ldots,s_{d+1}=0}^{p-1} \hat{g}_{\bar{\mathbf{s}}}^{\bar{A}\bar{B}}(x_1) e^{2\pi i \boldsymbol{\xi}_{\bar{\mathbf{s}}}^{\bar{B}} \boldsymbol{y}/N}$$

$$= \sum_{s_2,\ldots,s_{d+1}=0}^{p-1} \sum_{r_2,\ldots,r_{d+1}=0}^{p-1} ((\mathbf{M}^{\bar{A}\bar{B}})^{-1})_{\bar{\mathbf{s}};\bar{\mathbf{r}}} v_{x_1}(\boldsymbol{y}_{\bar{\mathbf{r}}}^{\bar{A}}) e^{2\pi i \boldsymbol{\xi}_{\bar{\mathbf{s}}}^{\bar{B}} \boldsymbol{y}/N}. \tag{4.56}$$

Using the interpolation in the first coordinate direction for fixed $s_2, \ldots, s_{d+1}$ implies

$$\mathcal{J}_p^{A_1 B_1} \hat{g}_{\bar{\mathbf{s}}}^{\bar{A}\bar{B}}(x_1) = \sum_{s_1=0}^{p-1} \left(\sum_{r_1=0}^{p-1} ((\mathbf{M}^{A_1 B_1})^{-1})_{s_1, r_1} \hat{g}_{\bar{\mathbf{s}}}^{\bar{A}\bar{B}}(x_{r_1}^{A_1}) \right) \mathrm{e}^{2\pi \mathrm{i}\xi_{s_1}^{B_1} x_1/N}. \tag{4.57}$$

Combining Equation (4.56) and (4.57) gives the new coefficients

$$\hat{g}_{s_1,\ldots,s_{d+1}}^{AB} := \sum_{r_1=0}^{p-1} ((\mathbf{M}^{A_1 B_1})^{-1})_{s_1, r_1} \hat{g}_{\bar{\mathbf{s}}}^{\bar{A}\bar{B}}(x_{r_1}^{A_1})$$

$$= \sum_{r_1,\ldots,r_{d+1}=0}^{p-1} ((\mathbf{M}^{A_1 B_1})^{-1})_{s_1, r_1} ((\mathbf{M}^{\bar{A}\bar{B}})^{-1})_{\bar{\mathbf{s}}, \bar{\mathbf{r}}} v_{x_{r_1}^{A_1}}(\mathbf{y}_{\bar{\mathbf{r}}}^{\bar{A}})$$

for the interpolating function

$$\mathcal{J}_p^{AB} g(\mathbf{x}) = \sum_{s_1,\ldots,s_d=0}^{p-1} \hat{g}_{s_1,\ldots,s_{d+1}}^{AB} (x_{r_1}^{A_1}) \mathrm{e}^{2\pi \mathrm{i}\boldsymbol{\xi}_{\mathbf{s}}^B \mathbf{x}/N}.$$

We have $\mathbf{M}^{A_1 B_1} \otimes \mathbf{M}^{\bar{A}\bar{B}} = \mathbf{M}^{AB}$ and $v_{x_{r_1}^{A_1}}(\mathbf{y}_{\bar{\mathbf{r}}}^{\bar{A}}) = g(\mathbf{x}_{\mathbf{r}}^A)$. Hence, the assertion follows with Lemma 2.3 (ii). $\qquad\square$

We use the same notation as in Theorem 4.20 and obtain the monomial representation

$$\mathcal{J}_p^{AB} g(\mathbf{x}) = \sum_{s_1,\ldots,s_d=0}^{p-1} \hat{g}_{s_1,\ldots,s_d}^{AB} \mathrm{e}^{2\pi \mathrm{i}\boldsymbol{\xi}_{\mathbf{s}}^B \mathbf{x}/N},$$

where $\hat{\mathbf{g}}^{AB} = (\mathbf{M}^{AB})^{-1} \mathbf{g}^{AB}$. Since $\mathbf{M}^{AB} = \bigotimes_{\nu=1}^d \mathbf{M}^{A_\nu B_\nu}$ and due to Equation (4.34) and Lemma 2.3 (i) we obtain with the matrices $\mathbf{D}_1^{A_\nu B_\nu}$, $\mathbf{D}_2^{A_\nu B_\nu}, \mathbf{G} \in \mathbb{C}^{p \times p}$ for the one-dimensional boxes A_ν, see Section 4.1.3.1, the decomposition

$$\mathbf{M}^{AB} = (\otimes_{\nu=1}^d \mathbf{D}_1^{A_\nu B_\nu})(\otimes_{\nu=1}^d \mathbf{G})(\otimes_{\nu=1}^d \mathbf{D}_2^{A_\nu B_\nu}).$$

Combining the previous equation with Equation (4.31) in the multi-dimensional case, it follows

$$\hat{\mathbf{f}}^{AB} = (\otimes_{\nu=1}^d (\mathbf{D}_2^{A_\nu B_\nu})^{-1})(\otimes_{\nu=1}^d \mathbf{G}^{-1})(\otimes_{\nu=1}^d (\mathbf{D}_1^{A_\nu B_\nu})^{-1}) \mathbf{f}^{AB}, \tag{4.58}$$

where $\mathbf{f}^{AB} = \sum_{S \in S_B} f^{PS}(\mathbf{x}_{\mathbf{r}}^A))_{r_1,\ldots,r_d=0}^{p-1}$. Given the coefficients

$$\hat{\mathbf{f}}^{PS} = (\hat{f}_{\mathbf{s}}^{PS})_{s_1,\ldots,s_d=0,\ldots,p-1} \in \mathbb{C}^{p^d},$$

we compute

$$\mathbf{f}^{AB} = \sum_{S \in S_B} \mathbf{N}^{AS} \hat{\mathbf{f}}^{PS}, \quad \mathbf{N}^{AS} := \left(\mathrm{e}^{2\pi \mathrm{i}\boldsymbol{\xi}_{\mathbf{s}}^S \mathbf{x}_{\mathbf{r}}^A/N} \right)_{r_1,\ldots,r_d; s_1,\ldots,s_d=0}^{p-1} = \otimes_{\nu=1}^d \mathbf{N}^{A_\nu B_\nu},$$

where $\mathbf{N}^{A_\nu B_\nu} \in \mathbb{C}^{p \times p}$ is the matrix of the one-dimensional case for the one-dimensional boxes A_ν and B_ν, see Section 4.1.3.1. We set $S = S_1 \times \cdots \times S_d$ for $S \in S_B$ and due to factorization (4.36) we finally obtain

$$\mathbf{f}^{AB} = \sum_{S \in S_B} (\otimes_{\nu=1}^{d} \mathbf{E}_1^{A_\nu S_\nu})(\otimes_{\nu=1}^{d} \mathbf{H})(\otimes_{\nu=1}^{d} \mathbf{E}_2^{A_\nu S_\nu}) \hat{\mathbf{f}}^{PS}.$$

Using the definition of $\mathbf{C}^{A_\nu S_\nu}$ for the one-dimensional boxes A_ν and S_ν in Section 4.1.3.1 the following relation holds true with $\mathbf{C}^{AS} := \otimes_{\nu=1}^{d} \mathbf{C}^{A_\nu S_\nu}$,

$$\hat{\mathbf{f}}^{AB} = (\otimes_{\nu=1}^{d} (\mathbf{D}_2^{A_\nu B_\nu})^{-1})(\otimes_{\nu=1}^{d} \mathbf{G}^{-1}) \sum_{S \in S_B} (\otimes_{\nu=1}^{d} \mathbf{C}^{A_\nu S_\nu})(\otimes_{\nu=1}^{d} \mathbf{H})(\otimes_{\nu=1}^{d} \mathbf{E}_2^{A_\nu S_\nu}) \hat{\mathbf{f}}^{PS}. \quad (4.59)$$

We denote by $\mathbf{d}_2^{A_\nu B_\nu}$, $\mathbf{c}^{A_\nu B_\nu}$, $\mathbf{e}_2^{A_\nu S_\nu} \in \mathbb{C}^p$ the vectors along the diagonal of the diagonal matrices $(\mathbf{D}_2^{A_\nu B_\nu})^{-1}$, $\mathbf{C}^{A_\nu B_\nu}$, $\mathbf{E}_2^{A_\nu S_\nu} \in \mathbb{C}^{p \times p}$, respectively. Furthermore, we declare the d-tensors $\mathcal{D}_2^{AB}$, $\mathcal{C}^{AS}$, $\mathcal{E}_2^{AS}$, $\hat{\mathcal{F}}^{PS} \in \mathbb{C}^{p \times \cdots \times p}$, such that

$$\operatorname{vec} \mathcal{D}_2^{AB} = \otimes_{\nu=1}^{d} \mathbf{d}_2^{A_\nu B_\nu}$$
$$\operatorname{vec} \mathcal{C}^{AS} = \otimes_{\nu=1}^{d} \mathbf{c}^{A_\nu S_\nu},$$
$$\operatorname{vec} \mathcal{E}_2^{AS} = \otimes_{\nu=1}^{d} \mathbf{e}_2^{A_\nu S_\nu},$$
$$\operatorname{vec} \hat{\mathcal{F}}^{PS} = \hat{\mathbf{f}}^{PS}.$$

Since $\otimes_{\nu=1}^{d} \mathbf{G}^{-1} = (\mathbf{G}^{-1} \otimes \mathbf{I}_{d-1})(\mathbf{I}_1 \otimes \mathbf{G}^{-1} \otimes \mathbf{I}_{d-2}) \ldots (\mathbf{I}_{d-1} \otimes \mathbf{G}^{-1})$, we can apply Lemma 2.7 (ii) and (iii) to obtain

$$\hat{\mathcal{F}}^{AB} = \left(\left(\sum_{S \in S_B} ((\mathcal{E}_2^{AS} \odot \hat{\mathcal{F}}^{PS}) \times_1 \mathbf{H} \cdots \times_d \mathbf{H}) \odot \mathcal{C}^{AS} \right) \right.$$
$$\left. \times_1 \mathbf{G}^{-1} \cdots \times_d \mathbf{G}^{-1} \right) \odot \mathcal{D}_2^{AB}, \quad (4.60)$$

where we set $\hat{\mathbf{f}}^{AB} := \operatorname{vec} \hat{\mathcal{F}}^{AB}$.

Remark 4.21. *(i) We apply Lemmata 2.7 (iv) and 2.3 (iv) in the two-dimensional case to Equation* (4.60) *and it reads with* $\hat{\mathbf{F}}^{PS} = \hat{\mathcal{F}}^{PS} \in \mathbb{C}^{p \times p}$ *as*

$$\tilde{\mathbf{F}}^{AB} := \sum_{S \in S_B} (\mathbf{c}^{A_2 S_2}(\mathbf{c}^{A_1 S_1})^\top) \odot \left(\mathbf{H} \left((\mathbf{e}_2^{A_2 S_2}(\mathbf{e}_2^{A_1 S_1})^\top) \odot \hat{\mathbf{F}}^{PS} \right) \mathbf{H}^\top \right),$$
$$\mathbf{f}^{AB} = \operatorname{vec} \left((\mathbf{d}_2^{A_2 B_2}(\mathbf{d}_2^{A_1 B_1})^\top) \odot \left(\mathbf{G}^{-1} \tilde{\mathbf{F}}^{AB} \mathbf{G}^{-\top} \right) \right). \quad (4.61)$$

(ii) Equation (4.58) *can be written with tensors as*

$$\hat{\mathbf{f}}^{AB} = \operatorname{vec}(((\mathcal{D}_1^{AB} \odot \mathcal{F}^{AB}) \times_1 \mathbf{G}^{-1} \cdots \times_d \mathbf{G}^{-1}) \odot \mathcal{D}_2^{AB}), \quad (4.62)$$

where $\mathbf{d}_1^{A_\nu B_\nu} \in \mathbb{C}^p$ *denote the vector along the diagonal of the diagonal matrix* $(\mathbf{D}_1^{A_\nu B_\nu})^{-1}$ *and the d-tensor* $\mathcal{D}_1^{AB} \in \mathbb{C}^{p \times \cdots \times p}$ *is defined such that*

$$\operatorname{vec} \mathcal{D}_1^{AB} = \otimes_{\nu=1}^{d} \mathbf{d}_1^{A_\nu B_\nu}.$$

4.2.3.2 Lagrange type basis

Let $g \in E_B(A)$. The interpolating function is given by Definition 4.17,

$$\left(\mathcal{J}_p^{AB}g\right)(\boldsymbol{x}) = d^B(\boldsymbol{x}) \sum_{r_1,\ldots,r_2=0}^{p-1} g(\boldsymbol{x}_r^A)d^B(-\boldsymbol{x}_r^A)\prod_{\nu=1}^{d} l_{r_\nu}^{A_\nu B_\nu}(x_\nu).$$

Combining the last equation with Equation (4.31) we obtain

$$f^{AB}(\boldsymbol{x}_{\mathbf{r}}^A) = \sum_{S\in S_B} d^S(\boldsymbol{x}_{\mathbf{r}}^A) \sum_{s_1,\ldots,s_d=0}^{p-1} f^{PS}(\boldsymbol{x}_{\mathbf{s}}^P)d^S(-\boldsymbol{x}_{\mathbf{s}}^P)\prod_{\nu=1}^{d} l_{s_\nu}^{P_\nu S_\nu}(\boldsymbol{x}_{\mathbf{r}}^A).$$

We define the matrices for higher dimensions with the matrices of the one-dimensional case as defined in Section 4.1.3.2,

$$\mathbf{R}^{AS} := \otimes_{\nu=1}^d \mathbf{R}^{A_\nu S_\nu}, \quad \mathbf{S}^{PS} := \otimes_{\nu=1}^d \mathbf{S}^{P_\nu S_\nu}, \quad \mathbf{L}^A := \otimes_{\nu=1}^d \mathbf{L}^{A_\nu}$$

with $\mathbf{L}^{A_\nu} := \left(l_s^{P_\nu S_\nu}(x_r^{A_\nu})\right)_{r,s=0}^{p-1}$. Altogether, we obtain

$$\mathbf{f}^{AB} = \sum_{S\in S_B} \mathbf{R}^{AS}\mathbf{L}^A\mathbf{S}^{AS}\mathbf{f}^{PS} = \sum_{S\in S_B} (\otimes_{\nu=1}^d \mathbf{R}^{A_\nu S_\nu})(\otimes_{\nu=1}^d \mathbf{L}^{A_\nu S_\nu})(\otimes_{\nu=1}^d \mathbf{S}^{P_\nu S_\nu})\mathbf{f}^{PS},$$

where $\mathbf{f}^{AB} := \left(f^{AB}(\boldsymbol{x}_{\mathbf{r}}^A)\right)_{r_1,\ldots,r_d=0,\ldots,p-1}$ are the function values of the current level and $\mathbf{f}^{PS} = \left(f^{PS}(\boldsymbol{x}_{\mathbf{s}}^P)\right)_{s_1,\ldots,s_d=0,\ldots,p-1}$ are the function values of the previous level. We denote by $\mathbf{s}^{P_\nu S_\nu}$, $\mathbf{r}^{A_\nu S_\nu}, \in \mathbb{C}^p$ the vectors along the diagonal of the diagonal matrices $\mathbf{S}^{P_\nu S_\nu}$, $\mathbf{R}^{A_\nu S_\nu}$, respectively. Moreover, we define the d-tensors $\mathcal{S}^{PS}, \mathcal{R}^{AS}, \mathcal{F}^{PS} \in \mathbb{C}^{p\times\cdots\times d}$ by

$$\text{vec}\,\mathcal{S}^{PS} = \otimes_{\nu=1}^d \mathbf{s}^{P_\nu S_\nu},$$
$$\text{vec}\,\mathcal{R}^{AS} = \otimes_{\nu=1}^d \mathbf{r}^{A_\nu S_\nu},$$
$$\text{vec}\,\mathcal{F}^{PS} = \mathbf{f}^{PS}.$$

Applying Lemma 2.7 leads to the representation

$$\mathbf{f}^{AB} = \text{vec}(\sum_{S\in S_B} \mathcal{R}^{AS} \odot ((\mathcal{S}^{PS} \odot \mathcal{F}^{PS}) \times_1 \mathbf{L}^{A_d}\cdots\times_d \mathbf{L}^{A_1})). \tag{4.63}$$

The computation in the two-dimensional case reads as follows

$$\mathbf{f}^{AB} = \text{vec}\left(\sum_{S\in S_B} \left(\mathbf{r}^{A_2 S_2}(\mathbf{r}^{A_1 S_1})^\top\right) \odot \left(\mathbf{L}^{A_2}\left((\mathbf{s}^{P_2 S_2}(\mathbf{s}^{P_1 S_1})^\top) \odot \mathbf{F}^{PS}\right)(\mathbf{L}^{A_1})^\top\right)\right).$$

4.3 Error analysis

In contrast to other analysis-based fast algorithms, the butterfly scheme uses a sequence of approximations and the local expansion degree depends not only on the target accuracy $\varepsilon > 0$ but also mildly on the nonharmonic bandwidth N. This behavior is illustrated also numerically in Section 4.5.2.

Theorem 4.22. *Let $L \in \mathbb{N}$, $N = 2^L$, $X, \Omega \subset [0, N]^d$, and $p \in \mathbb{N}$, $p \geq 5$, then the approximation (4.53) to the function (4.1) obeys the error estimate*

$$\|f - \widetilde{f}\|_{C(X)} \leq \frac{(C_p + 1)(C_p^{d(L+1)} - 1)}{C_p - 1} c_p \|\hat{\mathbf{f}}\|_1,$$

where C_p is given by Equation (4.15) and c_p by Equation (4.19).

Proof. Define for all levels $\ell = 0, \ldots, L$ and the frequency indices $\mathbf{n} \in \mathbb{N}_0^d$, $\|\mathbf{n}\|_\infty < 2^{L-\ell}$, the local sums

$$f^{\Omega_{L-\ell,\mathbf{n}}} := \sum_{\boldsymbol{\xi}_k \in \Omega_{L-\ell,\mathbf{n}} \cap \widetilde{\Omega}} \hat{f}_k e^{2\pi i \boldsymbol{\xi}_k \boldsymbol{x}/N}$$

and for the spatial indices $\mathbf{m} \in \mathbb{N}_0^d$, $\|\mathbf{m}\|_\infty < 2^\ell$, the error term

$$E_{\ell,\mathbf{m}} := \sum_{\mathbf{n} \in \mathbb{N}_0^d; \|\mathbf{n}\|_\infty < 2^{L-\ell}} \|f^{\Omega_{L-\ell,\mathbf{n}}} - f^{X_{\ell,\mathbf{m}} \Omega_{L-\ell,\mathbf{n}}}\|_{C(X_{\ell,\mathbf{m}})},$$

which by definition fulfills $\|f - \widetilde{f}\|_{C(X)} = \max_{\mathbf{m} \in \mathbb{N}_0^d, \|\mathbf{m}\|_\infty < 2^L} E_{L,\mathbf{m}}$. Using Theorem 4.19 and the triangle inequality, this quantity can be bounded for the zeroth level by

$$E_{0,0} = \sum_{\mathbf{n} \in \mathbb{N}_0^d; \|\mathbf{n}\|_\infty < 2^L} \|f^{\Omega_{L,\mathbf{n}}} - \mathcal{J}_p^{X_{0,0} T_{L,\mathbf{n}}} f^{\Omega_{L,\mathbf{n}}}\|_{C(X_{0,0})}$$

$$\leq \frac{(1 + C_p)(C_p^d - 1)}{C_p - 1} c_p \sum_{\mathbf{n} \in \mathbb{N}_0^d; \|\mathbf{n}\|_\infty < 2^L} \sum_{\boldsymbol{\xi}_k \in \Omega_{L,\mathbf{n}} \cap \widetilde{\Omega}} |\hat{f}_k|$$

$$\leq \frac{(1 + C_p)(C_p^d - 1)}{C_p - 1} c_p \|\hat{\mathbf{f}}\|_1.$$

For $\ell > 0$, adding and subtracting the term $\mathcal{J}_p^{X_{\ell,\mathbf{m}} \Omega_{L-\ell,\mathbf{n}}} f^{\Omega_{L-\ell,\mathbf{n}}}$ yields

$$E_{\ell,\mathbf{m}} = \sum_{\substack{\mathbf{n} \in \mathbb{N}_0^d \\ \|\mathbf{n}\|_\infty < 2^{L-\ell}}} \|f^{\Omega_{L-\ell,\mathbf{n}}} - \mathcal{J}_p^{X_{\ell,\mathbf{m}} \Omega_{L-\ell,\mathbf{n}}} f^{\Omega_{L-\ell,\mathbf{n}}} +$$

$$+ \mathcal{J}_p^{X_{\ell,\mathbf{m}} \Omega_{L-\ell,\mathbf{n}}} f^{\Omega_{L-\ell,\mathbf{n}}} - f^{X_{\ell,\mathbf{m}} \Omega_{L-\ell,\mathbf{n}}}\|_{C(X_{\ell,\mathbf{m}})}.$$

Using the triangle inequality, the first norm can be bounded as for the zeroth level, and we proceed by applying Equation (4.52), factoring out the interpolation operator, using the dyadic decomposition $\Omega_{L-\ell,\mathbf{n}} = \cup_{\mathbf{k} \in \{0,1\}^d} \Omega_{L-\ell+1,2\mathbf{n}+\mathbf{k}}$, and the relation $X_{\ell,\mathbf{m}} \subset X_{\ell-1,\lfloor \mathbf{m}/2 \rfloor}$ to obtain

$$\sum_{\substack{\mathbf{n} \in \mathbb{N}_0^d \\ \|\mathbf{n}\|_\infty < 2^{L-\ell}}} \|\mathcal{J}_p^{X_{\ell,\mathbf{m}} \Omega_{L-\ell,\mathbf{n}}} f^{\Omega_{L-\ell,\mathbf{n}}} - f^{X_{\ell,\mathbf{m}} \Omega_{L-\ell,\mathbf{n}}}\|_{C(X_{\ell,\mathbf{m}})}$$

$$\leq \sum_{\substack{\boldsymbol{n} \in \mathbb{N}_0^d \\ \|\boldsymbol{n}\|_\infty < 2^{L-\ell}}} \left\|\mathcal{J}_p^{X_{\ell,\mathbf{m}} \Omega_{L-\ell,\mathbf{n}}}\right\| \sum_{\mathbf{k} \in \{0,1\}^d} \left\|f^{\Omega_{L-\ell+1,2\mathbf{n}+\mathbf{k}}} - f^{X_{\ell-1,\lfloor \frac{\mathbf{m}}{2} \rfloor} \Omega_{L-\ell+1,2\mathbf{n}+\mathbf{k}}}\right\|_{C(X_{\ell-1,\lfloor \frac{\mathbf{m}}{2} \rfloor})}$$

$$\leq C_p^d E_{\ell-1,\lfloor \frac{\mathbf{m}}{2} \rfloor}.$$

Hence, we inductively find for $\mathbf{m} \in \mathbb{N}_0^d$, $\|\mathbf{m}\|_\infty < 2^L$, the relation

$$
\begin{aligned}
E_{L,\mathbf{m}} &\leq \frac{(1+C_p)(C_p^d-1)}{C_p-1} c_p \|\hat{\mathbf{f}}\|_1 + C_p^d E_{L-1,\lfloor \frac{\mathbf{m}}{2} \rfloor} \\
&\leq \frac{(1+C_p)(C_p^d-1)}{C_p-1} c_p \|\hat{\mathbf{f}}\|_1 + C_p^d \left(\frac{(1+C_p)(C_p^d-1)}{C_p-1} c_p \|\hat{\mathbf{f}}\|_1 + C_p^d E_{L-2,\lfloor \frac{\mathbf{m}}{4} \rfloor} \right) \\
&\leq \cdots \leq \frac{(1+C_p)(C_p^d-1)}{C_p-1} c_p \|\hat{\mathbf{f}}\|_1 \sum_{n=0}^{L} C_p^{nd} \\
&\leq \frac{(1+C_p)(C_p^d-1)}{C_p-1} c_p \cdot \frac{C_p^{d(L+1)}-1}{C_p^d-1} \|\hat{\mathbf{f}}\|_1,
\end{aligned}
$$

which proves the assertion. □

Corollary 4.23. *Under the assumptions of Theorem 4.22, let for given $\varepsilon \in (0,1]$ the expansion degree $p \in \mathbb{N}$ fulfill*

$$
p \geq \max \left\{ 10, 2 \left| \log \frac{2}{5}\varepsilon \right|, 2d(L+1) \right\},
$$

then $\|f - \widetilde{f}\|_{C(X)} \leq \varepsilon \|\hat{\mathbf{f}}\|_1$.

Proof. Direct calculation shows for $p \geq 10$ the relation

$$
p \geq 1 + \pi \sqrt{\mathrm{e}C_p}.
$$

Together with $p \geq 2d(L+1)$ this yields

$$
\log \frac{p-1}{\pi} - \frac{d(L+1)}{p} \log C_p \geq \frac{1}{2}
$$

and multiplication by $p \geq 2|\log \frac{2}{5}\varepsilon|$ finally gives the bound

$$
C_p^{d(L+1)} \left(\frac{\pi}{p-1} \right)^p \leq \frac{2}{5}\varepsilon
$$

Since $1 \leq K_p$, see Equation 4.16, we have

$$
1 + \frac{2}{\pi} \log p \leq C_p = K_p \left(1 + \frac{2}{\pi} \log p \right)
$$

and with $p \geq 10$ finally

$$
\frac{C_p+1}{C_p-1} = 1 + \frac{2}{C_p-1} \leq 1 + \frac{\pi}{\log p} \leq \frac{5}{2}.
$$

from which the assertion follows by Theorem 4.22. □

4.4 Complexity analysis

We consider the complexity of Algorithm 3 and 5 in due consideration of the computational realizations of Section 4.1.3 and 4.2.3 to compute the sum (4.1). The organization of the sampling nodes X, Ω is done in a precomputation step, where the trees $\mathcal{T}_X$ and $\mathcal{T}_\Omega$ are computed. Furthermore, we precompute the matrices $\mathbf{G}^{-1}$, $\mathbf{H}$, and $\mathbf{L}^A$, since $\mathbf{L}^A$ depends only on the relative position to the parent box P of A, cf. Equation (4.40).

4.4.1 The univariate case

We start with the univariate Algorithm 3. For the zeroth level in Step 2, the local sums $f^{\Omega_{L,n}}$ in (4.30) are evaluated at Chebyshev nodes x_r^X, $r = 0,\ldots,p-1$, which takes $\mathcal{O}(pM_2)$ floating point operations in total. In the monomial type basis approach, the coefficients $\hat{\mathbf{f}}^{X\Omega_{L,n}}$ for $n = 0,\ldots,2^L - 1$ are computed by Equation (4.35), where $\mathbf{f}^{X\Omega_{L,n}}$ is the vector of function values in the Chebyshev nodes. For Step 3 we have to compute the sums in (4.31). We have to apply the interpolation operator for each level $\ell = 0,\ldots,L$ and for at most N pairs in each level. A single application via the monomial type realization (4.37) or via the Lagrange type realization (4.42) takes $\mathcal{O}(p^2)$ floating point operations. Finally, we evaluate in Step 4 the function $\tilde{f}$ at all sampling nodes $x_j \in \tilde{X}$, which takes $\mathcal{O}(pM_1)$ floating point operations for the monomial type representation and for the Lagrange type representation with precomputation of the Lagrange functions $l_s^{X_{L,m}\Omega}$ at the evaluation nodes $x_j \in X_{L,m}$ for $s = 0,\ldots,p-1$. Without such precomputations, a straightforward evaluation of the Lagrange functions leads to $\mathcal{O}(p^2 M_1)$ floating point operations for this step. Since $L = \log_2 N$, this sums up to the total computational costs

- $\mathcal{O}(p(M_1 + M_2) + p^2 N \log N)$ with precomputed Lagrange functions or monomials,

- $\mathcal{O}(pM_2 + p^2(M_1 + N \log N))$ with directly evaluated Lagrange functions.

Assuming $M_1, M_2 = \mathcal{O}(N)$, a target accuracy $\varepsilon > 0$, and setting $p = C \log \frac{N}{\varepsilon}$, cf. Corollary 4.23, finally leads to the total computational costs

$$\mathcal{O}((|\log \varepsilon| + \log N)^2 N \log N). \tag{4.64}$$

Compared to [41], we use here the fact that the Lagrange matrices (4.41) are independent of the box A itself but depend only on the position to its father box P. This fact improves the time complexity by a factor p^3 for the Lagrange ansatz in the computation of the sums (4.42).

4.4.2 The multivariate case

Generalizing to $d \geq 2$, we assume that the sets $\tilde{X}, \tilde{\Omega} \subset [0,N]^d$ and their dyadic subdivisions are sparse in the sense

$$|\{\mathbf{m} \in \mathbb{N}_0^d : m_1,\ldots,m_d \leq 2^\ell - 1, \ \tilde{X} \cap X_{\ell,\mathbf{m}} \neq \emptyset\}| \leq C2^{(d-1)\ell},$$
$$|\{\mathbf{n} \in \mathbb{N}_0^d : n_1,\ldots,n_d \leq 2^{L-\ell} - 1, \ \tilde{\Omega} \cap \Omega_{L-\ell,\mathbf{m}} \neq \emptyset\}| \leq C2^{(d-1)(L-\ell)}$$

for some absolute constant $C > 0$. In particular, we set $M_1 = |\widetilde{X}| = \mathcal{O}(N^{d-1})$ and $M_2 = |\widetilde{\Omega}| = \mathcal{O}(N^{d-1})$. The condition above is satisfied if the sets lie on some smooth $(d-1)$-dimensional manifold in $[0, N]^d$, see also Figure 4.6 or [58, Fig. 5]. Under

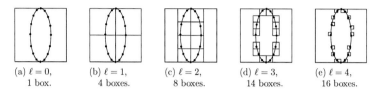

(a) $\ell = 0$,   (b) $\ell = 1$,   (c) $\ell = 2$,   (d) $\ell = 3$,   (e) $\ell = 4$,
1 box.   4 boxes.   8 boxes.   14 boxes.   16 boxes.

Figure 4.6: Example for $d = 2$. Nonempty boxes of samples on an ellipse.

this sparsity assumption, the number of admissible pairs for which we have to do computations is $\mathcal{O}(N^{d-1})$ in each level of the butterfly scheme. For the zeroth level in Algorithm 5 Step 2, the local sums $f^{\Omega_L, \mathbf{n}}$ in Equation (4.51) are evaluated in the Chebyshev nodes $\boldsymbol{x}_r^X$ for $\mathbf{r} \in \{0, \dots, p-1\}^d$, which takes $\mathcal{O}(p^d M_2)$ floating point operations. If we use the monomial type basis, we have to compute the coefficients $\hat{\mathbf{f}}^{X\Omega_L, \mathbf{n}}$ by Equation (4.62). Due to Lemma 2.7 (ii) a single k-mode product of $\mathbf{G}^{-1} \in \mathbb{C}^{p \times p}$ and a d-tensor $\mathcal{F} \in \mathbb{C}^{p \times \dots \times p}$, where the mode k-foldings of $\mathcal{F}$ are $\mathbf{F}_{(k)} \in \mathbb{C}^{p^{d-1} \times p}$ for $k = 1, \dots, d$, takes $\mathcal{O}(p^{d+1})$ floating point operations. In Step 3, the sums (4.31) have to be computed. Again, we have to apply the interpolation operator for each level $\ell = 0, \dots, L$ and for $\mathcal{O}(N^{d-1})$ pairs in each level. A single application via the monomial type realization (4.60) or via the Lagrange type realization (4.63) takes $\mathcal{O}(p^{d+1})$ operations. Finally in Step 4, the evaluation of the function $\widetilde{f}$ at all sampling nodes $\boldsymbol{x}_j \in \widetilde{X}$ takes $\mathcal{O}(p^d M_1)$ floating point operations in all approaches, in case of on-the-fly evaluations of the Lagrange functions in tensor product form the costs for the evaluation of the Lagrange functions are amortized. In total, this sums up to computational costs

$$\mathcal{O}(p^d(M_1 + M_2) + N^{d-1}p^{d+1} \log N). \tag{4.65}$$

Applying $M_1, M_2 = \mathcal{O}(N^{d-1})$ and $p = C \log \frac{N}{\varepsilon}$, cf. Corollary 4.23, leads to the total computational costs

$$\mathcal{O}(N^{d-1} \log N(|\log \varepsilon| + \log N)^{d+1}) \tag{4.66}$$

in Equation (4.65). The exponent $d+1$ of the last term can be decreased to d by using $\log c_p \leq -Cp \log p$ and the techniques in [16]. A similar improvement might be possible by reducing the p^d-dimensional ansatz to one which lives on the $(d-1)$-dimensional manifold. The butterfly algorithm can be parallelized, which reduces the complexity further, see [44].

4.4.3 Space complexity

Under the same assumptions as in the time complexity analysis, we discuss the space complexity of the BSFFT, see Algorithms 3 and 5 except the pre-computations in

Step 1. In each level $\ell \in \{0, \ldots, L\}$ of the butterfly scheme are N nodes (pairs of boxes (A, B)) in the univariate case and N^{d-1} in the multivariate case. The coefficient vectors $\hat{\mathbf{f}}^{AB} \in \mathbb{C}^{p^d}$ in the monomial representation or the vectors of function values $\mathbf{f}^{AB} \in \mathbb{C}^{p^d}$ in the Lagrange type realization, see Section 4.1.3 and 4.2.3, are stored. For all levels the space complexity is given by $\mathcal{O}(Np)$ for $d = 1$ and $\mathcal{O}(N^{d-1}p^d)$ for $d \geq 2$. Furthermore, the function values or coefficients of level $\ell \in \{1, \ldots, L\}$ are computed from the coefficients or function values from the previous level $\ell - 1$, see Equations (4.37), (4.42), for $d = 1$ and Equations (4.60), (4.63), for $d \geq 2$. Moreover, the matrices $\mathbf{G}^{-1}, \mathbf{H} \in \mathbb{C}^{p \times p}$, see Equation (4.34) and Equation (4.36), are stored in the monomial type realization. In the Lagrange type realization, the matrices $\mathbf{L}^A$, see Equation (4.41) are stored for all dimensions. The precomputation of the Lagrange polynomials evaluated in the samples $\widetilde{X}$ needs $|\widetilde{X}p|$ storage. Summing all over, the space complexity takes $\mathcal{O}(Np)$ in the univariate case and $\mathcal{O}(N^{d-1}p^d)$ in the multivariate case.

4.5 Numerical experiments

We use the same hard- and software as described in Section 3.4. The implementation of the BSFFT is realized for the dimensions $d = 1, 2, 3, 4$.

4.5.1 Local accuracy and stability

The first three experiments are dedicated to the approximation in one pair of admissible boxes. Let the nonharmonic bandwidth $N = 2^L$, $L = 10, 14$, the level $l = 5$, and the boxes $A = [0, 2^l]$, $B = [0, 2^{L-l}]$ be given. Draw nodes $x_j \in A$, $j = 1, \ldots, M_1$, $M_1 = N$, and $\xi_k \in B$, $k = 1, \ldots, M_2$, $M_2 = N$, at random from the uniform distribution and define the Fourier matrix

$$\mathbf{F} := \left(\mathrm{e}^{2\pi \mathrm{i}\xi_k \cdot x_j/N}\right)_{j=1,k=1}^{M_1,M_2}.$$

This matrix is approximately of low rank and we consider the Lagrange and monomial type expansions from Section 4.1.3,

$$\widetilde{\mathbf{F}} := \mathbf{F}^A (\mathbf{M}^{AB})^{-1}\mathbf{F}^B \quad \text{or} \quad \widetilde{\mathbf{F}} := \widetilde{\mathbf{L}}^A \mathbf{F}^B$$

with the auxiliary matrices

$$\mathbf{F}^A := \left(\mathrm{e}^{2\pi \mathrm{i}\xi_s^B x_j/N}\right)_{j=1,s=0}^{M_1,p-1}, \quad \mathbf{F}^B := \left(\mathrm{e}^{2\pi \mathrm{i}\xi_k x_r^A/N}\right)_{r=0,k=1}^{p-1,M_2},$$

$$\widetilde{\mathbf{L}}^A := \left(\mathrm{e}^{2\pi \mathrm{i}\left(c^B + \frac{w^B}{2}\right)x_j/N} l_r^{AB}(x_j)\mathrm{e}^{-2\pi \mathrm{i}\left(c^B + \frac{w^B}{2}\right)x_r^A/N}\right)_{j=1,r=0}^{M_1,p-1},$$

respectively. Due to Lemma 2.1 and Theorem 4.8, we obtain the local error

$$\max_{\substack{\hat{\mathbf{f}} \in \mathbb{C}^{M_2} \\ \|\hat{\mathbf{f}}\|_1=1}} \left\|\mathbf{F}\hat{\mathbf{f}} - \widetilde{\mathbf{F}}\hat{\mathbf{f}}\right\|_\infty = \|\mathbf{F} - \widetilde{\mathbf{F}}\|_{1\to\infty} \leq (1+C_p)c_p = \frac{(1+C_p)}{\pi p}\left(\frac{\pi}{p-1}\right)^p.$$

We compare the quantity $\|\mathbf{F} - \widetilde{\mathbf{F}}\|_{1\to\infty}$ for both realizations and for the original approach [58] in Figure 4.7. The original scheme differs from the variant of the monomial type basis approach in Section 4.1.3 in the choice of interpolation nodes in $x_r^A \in A$, where we use zeros of Chebyshev polynomials instead of extrema, and in the choice of the 'equivalent sources' $\xi_s^B \in B$, where we use equidistant points instead of Chebyshev extrema. Note, that an explicit representation for the interpolation in Chebyshev nodes in A and B is given in a Lagrange type basis in Definition 4.12 and an error estimate is given in Corollary 4.13. The original approach, the interpolation in Chebyshev extremal points in A and B, has the same convergence rate, cf. [53, Chapter 7 and 8]. In all cases, the error decays exponentially with increasing expansion degree p, as predicted by Theorem 4.8 for the equidistant points $\xi_s^B \in B$, see Figure 4.7. However, note that both monomial type approaches achieve only an accuracy $\|\mathbf{F} - \widetilde{\mathbf{F}}\|_{1\to\infty} \approx 10^{-8}$ and suffer from severe instabilities for values $p \geq 9$ which is well predicted by the quantity $\kappa(\mathbf{M}^{AB})\mu$, where $\mu \approx 2 \cdot 10^{-16}$ denotes the machine precision and κ the spectral condition number bounds, cf. Theorems 4.15 and 4.16. The second experiment, cf. Figure 4.8(a), illustrates the local error $\|\mathbf{F} - \widetilde{\mathbf{F}}\|_{1\to\infty}$ of the BSFFT in comparison to the singular value decomposition of the Fourier matrix $\mathbf{F}$ and Ying's variant, where we use now a Lagrange type realization for the interpolation in time and frequency domain in Chebyshev nodes, cf. Definition 4.12. The third experiment analyzes the stability of the monomial and the Lagrange type approaches as theoretically discussed in Section 4.1.4. Figure 4.8(b) shows the growth of the condition numbers of the matrices $\mathbf{M}^{AB}, \mathbf{L}^A \in \mathbb{C}^{p \times p}$, and their lower and upper bound, see Equations (4.43) and (4.46), respectively.

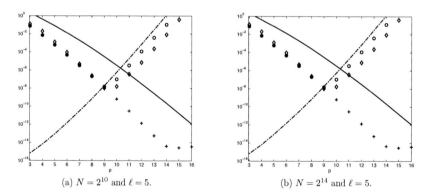

(a) $N = 2^{10}$ and $\ell = 5$.   (b) $N = 2^{14}$ and $\ell = 5$.

Figure 4.7: Relative error $\|\mathbf{F} - \widetilde{\mathbf{F}}\|_{1\to\infty}$ with respect to the local expansion degree p for the realization via Lagrange functions (plus), via monomials with equispaced frequencies (diamonds), and via monomials with Chebyshev extrema (circle). The theoretical estimate of Theorem 4.8 (solid line) and the quantity $\kappa(\mathbf{M}^{AB})\mu$ (dash-dot line) are illustrated, where $\mu \approx 2 \cdot 10^{-16}$ denotes the machine precision.

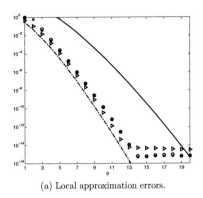

(a) Local approximation errors.

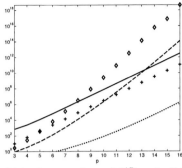

(b) Condition numbers $\kappa(\mathbf{M}^{AB})$ and $\kappa(\mathbf{L}^A)$ with their bounds.

Figure 4.8: Figure 4.8(a) illustrates the ℓ^2-approximation error (daggers) of the trigonometric interpolation, see Definition 4.1, the ℓ^2-approximation error of Ying's variant [58] (circle) using interpolation in both variables in Chebyshev nodes via representation of Definition 4.12 together with its normalized theoretical upper bound (solid line) given in Corollary 4.13 for $M_1 = M_2 = N$, and the singular values (triangles), representing the best approximation in this case, together with the lower bound of Remark 4.14 (dashed line). Figure 4.8(b) illustrates the condition number of the Vandermonde matrix $\mathbf{M}^{AB} \in \mathbb{C}^{p \times p}$ (diamonds) with their lower bounds (4.43) (dotted line) and (4.44) (dashed line), and the Lagrange matrix $\mathbf{L}^A \in \mathbb{C}^{p \times p}$ (plus) with their upper bound (4.46) (solid line) with respect to the local expansion degree p.

4.5.2 Accuracy

Regarding the accuracy of the whole algorithm, we draw coefficients

$$\hat{f}_k \in [-\frac{1}{2}, \frac{1}{2}] \times [-\frac{1}{2}, \frac{1}{2}]\mathrm{i}, \quad k = 1, \dots, M_2,$$

at random from the uniform distribution and consider the relative error as defined in Equation (3.9),

$$\varepsilon_1 = \frac{\|\mathbf{f} - \widetilde{\mathbf{f}}\|_\infty}{\|\hat{\mathbf{f}}\|_1} \leq \frac{(C_p + 1)(C_p^{d(L+1)} - 1)}{C_p - 1} c_p,$$

where $\mathbf{f} \in \mathbb{C}^{M_1}$ denotes the function values $f(\boldsymbol{x}_j)$ to evaluate the sums (4.1) and its approximation $\widetilde{\mathbf{f}} = (\widetilde{f}(\boldsymbol{x}_j))_{j=1,\dots,M_1}$, cf. Equation (4.53). For $d = 1, 2, 3$ we choose random sampling nodes X and $\widetilde{\Omega}$ from the uniform distribution. We set $M_1 = 1$ for $d = 1$ and $M_1 = N^{d-1}$ otherwise. In particular we choose the nodes on ellipses for $d = 2$ and on ellipsoids for $d = 3$. For $d = 4$, we choose equispaced nodes on a three-dimensional hyperplane.

We compare the quantity ε_1 and the upper bound from Theorem 4.22 for the monomial and the Lagrange type realization in Figure 4.9 (top). In these two tests for different parameters N as well as further experiments for $d = 2$, $d = 3$, and $d = 4$, the total error decays exponentially with p but is again limited for the monomial type realization, cf. Figures 4.9 and 4.10. In all cases, a least squares fit reveals a numerical error decay $\varepsilon_1 \approx C \cdot 16^{-p}$, where the constant C seems to depend neither on d nor L.

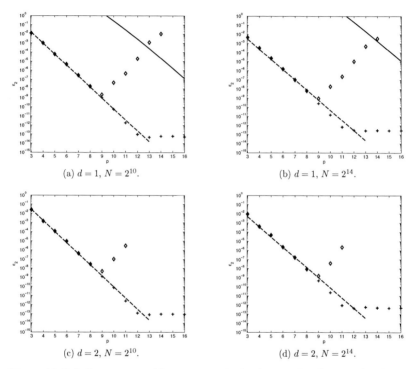

(a) $d = 1$, $N = 2^{10}$.

(b) $d = 1$, $N = 2^{14}$.

(c) $d = 2$, $N = 2^{10}$.

(d) $d = 2$, $N = 2^{14}$.

Figure 4.9: Relative error ε_1 with respect to the local expansion degree p for the realization via Lagrange functions (diamonds) with a least squares fit of the Lagrange type data (dotted line), via the monomials (plus) and the upper bound of Theorem 4.22 (solid line) for $d = 1, 2$.

Our second experiment touches the question whether the error really increases for increasing nonharmonic bandwidth as predicted by Corollary 4.23, i.e., $\varepsilon_1 \approx C_{p,d} N$. While randomly drawn coefficients $\hat{f}_k \in \mathbb{C}$, as in the previous test, do not show this increase, using constant coefficients $\hat{f}_k = 1$ support a weak increase

$$\varepsilon_1 \approx C_{p,d} L = C_{p,d} \log_2 N,$$

cf. Figure 4.11.

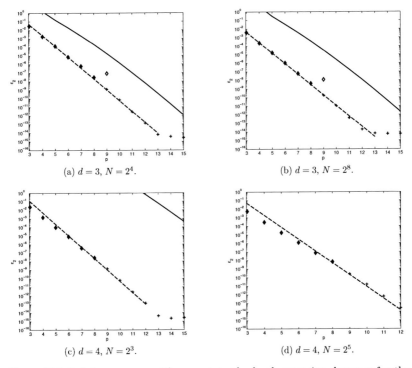

Figure 4.10: Relative error ε_1 with respect to the local expansion degree p for the realization via Lagrange functions (diamonds) with a least squares fit of the Lagrange type data (dotted line), via the monomials (plus) and the upper bound of Theorem 4.22 (solid line) for $d = 3, 4$.

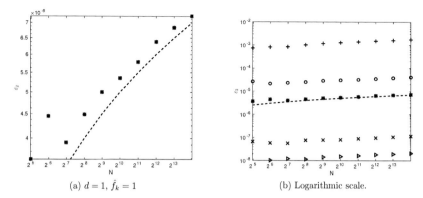

(a) $d = 1$, $\hat{f}_k = 1$ (b) Logarithmic scale.

Figure 4.11: The relative error ε_1 for $p = 5, 6, 7, 8, 9$ (plus, circle, asterisk, cross, right-pointing triangle) with respect to the nonharmonic bandwidth L and the constant $\frac{1}{2} \cdot 10^{-6}$ (dashed line).

4.5.3 Computational times

Finally, we compare the computational times, measured by the MATLAB functions `tic` and `toc`, of the naive evaluation (4.1) and Algorithm 3 or 5 for fixed spatial dimensions $d = 1, 2, 3, 4$, fixed local expansion degrees $p = 4, 8$, and with respect to increasing nonharmonic bandwidth N. We draw coefficients $\hat{f}_k \in [-\frac{1}{2}, \frac{1}{2}] \times [-\frac{1}{2}, \frac{1}{2}]$i and source and target nodes $\boldsymbol{\xi}_k, \boldsymbol{x}_j \in [0, N]^d$, $k = 1, \ldots, M_2$, $j = 1, \ldots, M_1$, at random from the uniform distribution (on the submanifold). As discussed in Section 4.1.3 and 4.2.3, we consider three realizations of the interpolation operator, the monomial type basis and the Lagrange type basis with and without pre-computations of the Lagrange functions at the final evaluation nodes. We compare these approximate variants with the naive computation of sums (4.1) given by Algorithm 6.

Figure 4.12 shows the measured times for $d = 1, 2$ and $M_1 = M_2 = N$ sampling nodes $\boldsymbol{\xi}_k, x_j \in [0, N]$, where the nodes lie on ellipses for $d = 2$ as described in the previous section. The break even with the naive method occurs at $N = 32$ for $d = 1$, and $N = 1024$ for $d = 2$. For $d = 1$, we see some increase for a larger local expansion degree $p = 8$ when no precomputation is done. Figure 4.13 gives the results for $d = 3, 4$ with $M_1 = M_2 = N^{d-1}$ sampling nodes on spheres and hyperplanes, respectively. As already mentioned in Section 4.4, the precomputation of the Lagrange functions in the last step of the algorithm does not gain any improvement here, cf. Figure 4.12. Finally, note that the break even with the naive algorithm occurs at a suitable problem size but a further reduction in absolute computing time is necessary for real applications.

In the next experiment, we compare the NNFFT (nonequispaced fast Fourier transform in time and frequency domain), see [15, 48] with the BSFFT. After some shifts, the

---

Algorithm 6 Naive Algorithm for Fourier transforms.

Input:

$d, L, M_1, M_2, p \in \mathbb{N}, \ p \geq 2, \ N = 2^L,$
$\hat{f}_k \in \mathbb{C}, \ k = 1, \ldots, M_2,$
$\widehat{\Omega} = \{ \boldsymbol{\xi}_k \in [0, N]^d : k = 1, \ldots, M_2 \},$
$\widetilde{X} = \{ \boldsymbol{x}_j \in [0, N]^d : j = 1, \ldots, M_1 \}.$

$f_j = 0, \ j = 1, \ldots, M_1$
for $j = 1, \ldots, M_1$ **do**
    for $k = 1, \ldots, M_2$ **do**
        $f_j = f_j + \mathrm{e}^{2\pi \mathrm{i} \boldsymbol{x}_j \boldsymbol{\xi}_k / N} \hat{f}_k$
    end for
end for

Output:
    Function values $f_j := f(\boldsymbol{x}_j), \ j = 1, \ldots, M_1.$

NNFFT computes approximate sums of the form

$$u(\boldsymbol{x}_j) = \sum_{k=1}^{M_2} \hat{u}_k \mathrm{e}^{2\pi \mathrm{i} \boldsymbol{\xi}_k \boldsymbol{x}_j / N} \quad \text{for } j = 1, \ldots, M_1,$$

where $N \in \mathbb{N}$ denotes the bandwidth and $\boldsymbol{x}_j, \boldsymbol{\xi}_k \in [0, N]^d$. This is for $d = 1$ the same as computing a butterfly transform (4.1). Computing a NNFFT takes $\mathcal{O}(N^d \log N)$ floating point operations. The NNFFT does not account the sparsity of the sampling nodes like the BSFFT. We compare the running times of both algorithms for $d = 1, 2$, which are illustrated in Figure 4.14. Both algorithms, the NNFFT and BSFFT are of order $\mathcal{O}(N \log N)$ for $d = 1$ but the constant of the NNFFT is much smaller than the constant of the BSFFT. The BSFFT takes $\mathcal{O}(N \log N)$ floating point operations in the two-dimensional case for sampling nodes on a smooth curve whereas the NNFFT takes $\mathcal{O}(N^2 \log N)$ floating point operations. Figure 4.14(b) illustrates the computational times for $d = 2$ and sampling nodes on ellipses. Since the constant of the BSFFT is large we obtain a break even for $N = 2^{13}$. Note that during the NNFFT an FFT for $d = 2$ of size $N \times N$ is performed. Hence, the NNFFT requires a high amount of memory, which makes the NNFFT not applicable in comparison to the BSFFT for large sparse data sets $\widetilde{X}$ and $\widehat{\Omega}$.

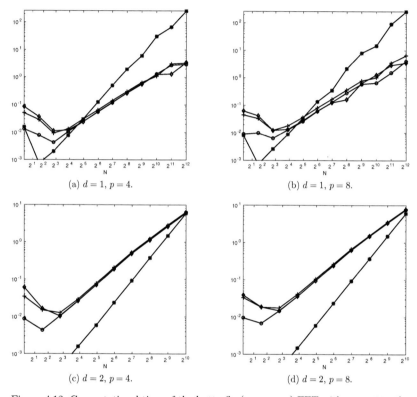

Figure 4.12: Computational time of the butterfly (nonsparse) FFT with respect to the nonharmonic bandwidth and problem size $N = M_1 = M_2$ via Lagrange type without pre-computations (plus), with pre-computations (circle), via Monomial type (diamonds) representation and the exact naive variant (asterisk) for $d = 1, 2$.

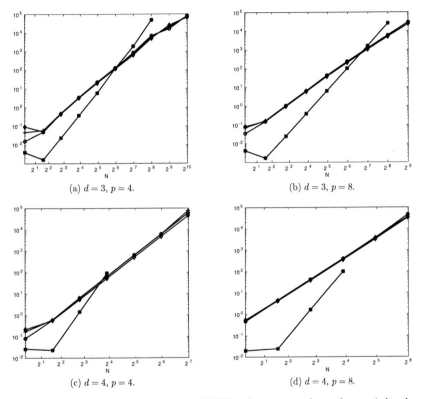

Figure 4.13: Computational time of the BSFFT with respect to the nonharmonic bandwidth N and problem sizes $M_1 = M_2 = N^{d-1}$ via Lagrange type without
pre-computations (plus), with pre-computations (circle), via Monomial
type representation (diamonds) and the exact naive variant (asterisk) for
$d = 3, 4$.

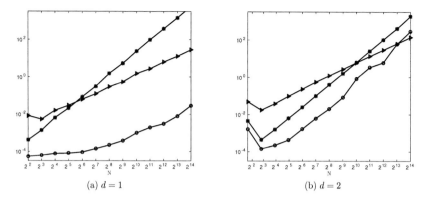

(a) $d = 1$ (b) $d = 2$

Figure 4.14: Computational times of the BSFFT for approximation rank $p = 8$ (right-pointing triangle), the NNFFT (circle) and the naive computation (asterisk) for the dimensions $d = 1, 2$ with respect to the problem sizes $M_1 = M_2 = N$.

4.5.4 Vibrating string problem as a simple application

There are several applications in modern imaging, where Fast Fourier transforms play an important role. For example in photo acoustic tomography, where spherical means are a suitable model for the solution of the three-dimensional wave equatiom. For example, in [24, 23] the authors use a four-dimensional BSFFT to compute spherical means of three-dimensional objects, see [24, Algorithm 2]. An application is also given in [58] by a scatterd solution of the Helmholtz-Equation. For simplicity, we treat a one-dimensional example, the one-dimensional wave equation. The solution of the one-dimensional wave equation can be computed by applying a two-dimensional BSFFT. We are looking for a function $u : [0,1] \times [0,T] \to \mathbb{R}$, $u \in C^2([0,1] \times [0,T])$, which fulfills the wave equation with initial values

$$
\begin{aligned}
\partial_t^2 u(x,t) &= c^2 \partial_x^2 u(x,t) \\
u(x,0) &= f(x), \qquad \partial_t u(x,0) = 0, \text{ for } x \in [0,1],
\end{aligned}
\tag{4.67}
$$

where $c > 0$ denotes a density constant and $f \in C([0,1])$. This problem is also known as vibrating string problem. For this simple one-dimensional situation, it can be proven, that the solution is given explicitly by

$$
u(x,t) = \frac{1}{2} \left(f(x+ct) + f(x-ct) \right).
$$

Moreover, the solution can be expressed as a series.

Lemma 4.24. *Let* $f \colon [0,1] \to \mathbb{R}$, $f \in C^2([0,1])$, *a 1-periodic function with Fourier coefficients*

$$
c_k(f) := \int_0^1 f(x) \mathrm{e}^{-2\pi \mathrm{i} k x} \mathrm{d}x, \quad k \in \mathbb{Z}.
$$

The solution of (4.67) is unique and given by

$$u(x,t) = \frac{1}{2} \sum_{k \in \mathbb{Z}} c_k(f) \left(e^{2\pi i (x+ct)k} + e^{2\pi i (x-ct)k} \right). \tag{4.68}$$

The series is absolutely and uniformly convergent.

Proof. See [7, section 7.1.3] for a proof. $\square$

We truncate the series (4.68) after N summands. Moreover, we approximate the Fourier coefficients $c_k(f)$ by the discrete Fourier coefficients $\hat{f}_k$,

$$c_k(f) = \frac{1}{N} \int_0^1 f(x) e^{-2\pi i k x} \mathrm{d}x \approx \frac{1}{N} \sum_{n=0}^{N-1} f\left(\frac{n}{N}\right) e^{-2\pi i k n/N} =: \hat{f}_k,$$

i.e. we compute approximatively the integral with the trapezoidal rule. Let $c = 1$, an approximation of the series (4.68) is given by

$$u(x,t) \approx \frac{1}{2} \sum_{k=-N/2}^{N/2-1} \hat{f}_k \left(e^{2\pi i k(x+t)} + e^{2\pi i k(x-t)} \right) := g(x,t). \tag{4.69}$$

The function $g \colon [0,1]^2 \to \mathbb{C}$ can be reformulated as

$$g(x,t) = \frac{1}{2} \sum_{k=-N/2}^{N/2-1} \hat{f}_k \left(e^{2\pi i (k,k)(x,t)^\top} + e^{2\pi i (k,-k)(x,t)^\top} \right)$$

$$= \frac{1}{2} e^{-\pi i (x,t)(N,N)^\top} \sum_{k=-\frac{N}{2}}^{\frac{N}{2}-1} \hat{f}_k \left(e^{2\pi i (x,t) \cdot (k+N/2, k+N/2)} + e^{2\pi i (x,t) \cdot (k+N/2, -k+N/2)} \right).$$

Let

$$\boldsymbol{x} := N(x,t) \in [0,N]^2,$$

$$\widetilde{\Omega} := \left\{ \boldsymbol{\xi} := (\xi_1, \xi_2) \in [0,N]^2 : \xi_1 = j + \frac{N}{2}, \quad \xi_2 = \pm j + \frac{N}{2} \text{ for } j = -\frac{N}{2}, \dots, \frac{N}{2} - 1 \right\},$$

$$\hat{g}_j := \begin{cases} 2\hat{f}_{(\boldsymbol{\xi}_j)_1 - \frac{N}{2}}, & \text{for } (\boldsymbol{\xi}_j)_1 = \frac{N}{2}, \\ \hat{f}_{(\boldsymbol{\xi}_j)_1 - \frac{N}{2}} & \text{otherwise} \end{cases}$$

and finally we obtain a two-dimensional BSFFT

$$g(\boldsymbol{x}) = \frac{1}{2} e^{-\pi i \boldsymbol{x} \cdot (N,N)/N} \sum_{\boldsymbol{\xi}_j \in \widetilde{\Omega}} \hat{g}_j e^{2\pi i \boldsymbol{x} \cdot \boldsymbol{\xi}_j / N}.$$

For an example, we choose a function $\tilde{f} \colon \mathbb{R} \to \mathbb{R}$,

$$\tilde{f}(x) := \begin{cases} \left(1 - 4\left(x - \frac{1}{2}\right)^2 \right)^2 & \text{for } x \in [0,1) \\ 0 & \text{otherwise.} \end{cases}$$

We consider the 1-periodic continuation f of $\tilde{f}$,

$$f(x) := \sum_{r \in \mathbb{Z}} \tilde{f}(x+r) \qquad \text{with Fourier coefficients}$$

$$c_k(f) = \begin{cases} \frac{8}{15} & \text{for } k=0 \\ -\frac{24}{\pi^4 k^4} & \text{otherwise.} \end{cases}$$

Since $\sum\limits_{k=1}^{\infty} \frac{1}{k^4} = \frac{\pi^4}{90}$, the Fourier coefficients obey $c_k(f) \in \ell^1$,

$$\sum_{k \in \mathbb{Z}} |c_k(f)| = \sum_{k=-\infty}^{1} \left| \frac{-24}{\pi^4 k^4} \right| + \frac{8}{15} + \sum_{k=1}^{\infty} \left| \frac{-24}{\pi^4 k^4} \right| = \frac{48}{\pi^4} \sum_{k=1}^{\infty} \frac{1}{k^4} + \frac{8}{15} = \frac{16}{15}.$$

Hence, the Fourier series of f converges absolutely and uniformly to f. Let $N = 2^L$, $L \in \mathbb{N}$, and $M = 1024$ be fixed. We want to compute the solution of the one-dimensional wave equation for a fixed spatial point $x_i = \frac{1}{2}$ at different times $t_i = \frac{i}{M}$ for $i = 0, \ldots, M-1$. We have to compute the discrete Fourier coefficients $\hat{\mathbf{f}} \in \mathbb{C}^N$ of $\mathbf{f} := (f(j/N))_{j=0,\ldots,N-1}$. Due to the periodicity of the coefficients $\hat{f}_k = \hat{f}_{k+rN}$, $r \in \mathbb{Z}$, we have

$$\hat{\mathbf{g}} = (\hat{f}_{N/2}, \ldots, \hat{f}_{N-1}, \hat{f}_0, \ldots, \hat{f}_{N/2-1})^\top.$$

The spatial and frequency nodes, the test function, the exact solution and the approximate solution by applying the BSFFT are illustrated in Figure 4.15. The absolute error $\max\limits_{i=0,\ldots,M-1} |u(x_i, t_i) - g(x_i, t_i)|$ is listed in Table 4.1 for different bandwidths N and approximation ranks p. On one hand, the error depends on the parameter N in the truncation, cf. Equation (4.69), and on the other hand on the approximation rank p in the BSFFT. Table 4.1 and Figure 4.16 illustrate this dependence of the absolute error on N and p.

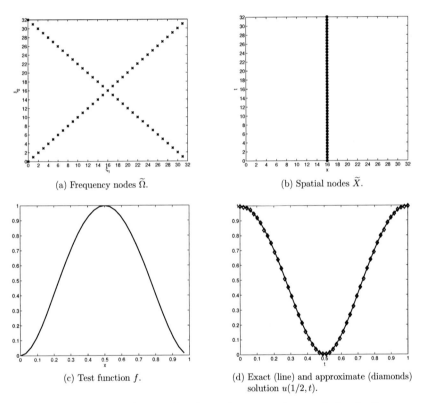

(a) Frequency nodes $\widetilde{\Omega}$.

(b) Spatial nodes $\widetilde{X}$.

(c) Test function f.

(d) Exact (line) and approximate (diamonds) solution $u(1/2, t)$.

Figure 4.15: Sampling nodes, test function and an approximate solution for the wave equation for $N = 32$, $M = 50$, and $p = 4$.

N	p	$\max\limits_{i=0,\ldots,M-1} \lvert u(x_i, t_i) - \tilde{g}(x_i, t_i) \rvert$
32	3	$9.1611 \cdot 10^{-2}$
	5	$8.0644 \cdot 10^{-4}$
	7	$3.3804 \cdot 10^{-5}$
	9	$3.3623 \cdot 10^{-5}$
256	3	$1.2123 \cdot 10^{-1}$
	5	$1.4154 \cdot 10^{-3}$
	7	$8.1478 \cdot 10^{-6}$
	9	$7.9908 \cdot 10^{-8}$
4096	3	$1.1892 \cdot 10^{-1}$
	5	$1.5176 \cdot 10^{-3}$
	7	$9.0309 \cdot 10^{-6}$
	9	$3.4970 \cdot 10^{-8}$

Table 4.1: Absolute errors of the approximate solution computed with the two-dimensional BSFFT of the wave equation for different pairs N, p, and a fixed number $M = 1024$ of spatial nodes.

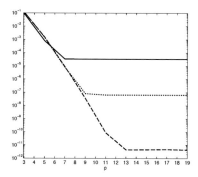

Figure 4.16: Absolute error of the approximate solution computed with the two-dimensional BSFFT of the wave equation for a fixed number $M = 1024$ of spatial nodes for the bandwidths $N = 32$ (solid), $N = 256$ (dotted), $N = 4096$ (dashed) in dependence of the approximation rank p.

Chapter 5

Fast Fourier transform for nonequispaced complex nodes

The BSFFT neither allows for an evaluation in complex nodes nor for complex frequencies. This issue has been already studied in [1] which allows for the computation of exponential sums where the complex evaluation nodes lie in a small strip around the real line. Here, we present two methods for the computation of sums

$$f_j := f(w_j) = \sum_{k=1}^{M_2} \hat{f}_k e^{2\pi i w_j \xi_k / N} \tag{5.1}$$

for $j = 1, \ldots, M_1$ with $\xi_k \in [0, N]$ and $w_j \in \mathbb{C}$. One method is a generalization of the BSFFT which also allows for complex evaluation nodes in a strip around the real line. The second method is based on a combination of the Laplace transform, see Section 3.2, with fast Fourier transforms, i.e., the NFFT, NNFFT, or BSFFT. The first question is, which restrictions on the complex spatial nodes $w_j \in \mathbb{C}$ are needed, to obtain a good low rank approximation of the kernel function $\kappa(w, \xi) := e^{2\pi i w \xi / N}$.

5.1 Generalization of the butterfly scheme to complex nodes in a strip

A first idea for the computation of the sum (5.1) is to expand the Definition 4.1 for boxes $D \subset \mathbb{C}$ instead of $A \subset \mathbb{R}$, where D is a small strip around the real line, which depends on the approximation rank. The following results were obtained in the supervised Bachelor-Thesis [20]. At first, we generalize Lemma 4.7 for complex values in a sector of an annulus to obtain a low rank approximation of the kernel κ on a small strip around the real line.

Lemma 5.1. *Let* $p \geq 5$, $p \in \mathbb{N}$, $\alpha \in [0, p-1]$,

$$\tilde{D}_p := \left[-\frac{1}{2}, \frac{1}{2}\right] \times i \frac{p-1}{2\pi} \left[\log(1 - \frac{1}{p}), \log(1 + \frac{1}{p})\right], \tag{5.2}$$

and $\tilde{\Gamma}_p := z(\tilde{D}_p)$, $z(y) := \mathrm{e}^{-2\pi \mathrm{i}\frac{y}{p-1}}$, where the map z assigns the complex strip $\tilde{D}_p$ to a sector on an annulus, see Figure 5.1. It follows

$$\left| z^\alpha - \sum_{k=0}^{p-1} \binom{\alpha}{k}(z-1)^k \right| \le \tilde{c}_p,$$

where

$$\tilde{c}_p := \frac{1}{\pi(p-1)}\left(\left(\frac{\pi}{p-1}\right)^2 \left(1+\frac{1}{p}\right) + \frac{1}{p^2}\right)^{p/2}. \tag{5.3}$$

Proof. Let $y \in \tilde{D}_p$ and $z = z(y) \in \tilde{\Gamma}_p$, then we have

$$|z-1|^2 = \left| \mathrm{e}^{\frac{2\pi\mathrm{i}}{p-1}(\mathrm{i}\,\mathrm{Im}(y)+\mathrm{Re}(y))} \right| \le \left| \mathrm{e}^{\frac{2\pi\mathrm{i}}{p-1}\max_{y\in\hat{C}}\,\mathrm{Im}(y)} \mathrm{e}^{\frac{2\pi\mathrm{i}}{p-1}\max_{y\in\hat{C}}\,\mathrm{Re}(y)} \right|$$

$$= \left| \left(1+\frac{1}{p}\right)\mathrm{e}^{-\frac{\pi\mathrm{i}}{p-1}} - 1\right|^2 = \left(1+\frac{1}{p}\right)^2 - 2\left(1+\frac{1}{p}\right)\cos\frac{\pi}{p-1} + 1$$

$$= 2(1-\cos\frac{\pi}{p-1})\left(1+\frac{1}{p}\right) + \frac{1}{p^2}.$$

The value, which maximizes $\max_{z\in\tilde{\Gamma}_p}|z-1|$ is denoted by z^* in Figure 5.1. Applying Equation (4.10) yields

$$|z-1|^2 \le \left(\frac{\pi}{p-1}\right)^2 \left(1+\frac{1}{p}\right) + \frac{1}{p^2}.$$

The rest of the proof is similar to the one of Lemma 4.7. Using $|z| \ge 1 - \frac{1}{p}$ in Equation (4.21) and applying Lemma 4.5 yields the assertion. $\qquad\square$

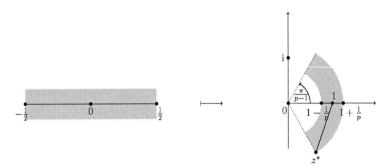

Figure 5.1: The map $\tilde{D}_p \mapsto \tilde{\Gamma}_p$.

We extend Definition 4.1 and 4.2 to a complex strip D. For simplicity of notation, we use the same symbols for the extended operators.

Definition 5.2. *For boxes $D \subset \mathbb{C}$, $D = A \times \mathrm{i}C$, $A, B, C \subset \mathbb{R}$, we define the linear space of all finite expansions of exponential functions*

$$E_B(D) := \left\{ g \colon D \to \mathbb{C} : g(x) = \sum_{j=1}^{K} \hat{g}_j \mathrm{e}^{2\pi\mathrm{i}\xi_j x/N}, \ K \in \mathbb{N}, \ \hat{g}_j \in \mathbb{C}, \ \xi_j \in B \right\},$$

and the subspace with $p \in \mathbb{N}$ equispaced frequencies

$$E_B^p(D) := \left\{ g \colon D \to \mathbb{C} : g(x) = \sum_{j=0}^{p-1} \hat{g}_j^{AB} \mathrm{e}^{2\pi\mathrm{i}\xi_j^B x/N}, \ \hat{g}_j^{AB} \in \mathbb{C} \right\}.$$

Moreover, we define for $p \in \mathbb{N}$, $p \geq 2$, the trigonometric interpolation operator

$$\mathcal{J}_p^{DB} \colon C(D) \to E_B^p(D), \quad g \mapsto \mathcal{J}_p^{DB} g = \sum_{s=0}^{p-1} \hat{g}_s^{DB} \mathrm{e}^{2\pi\mathrm{i}x\xi_s^B/N}, \tag{5.4}$$

such that in the Chebyshev nodes x_r^A the interpolation condition $\mathcal{J}_p^{DB} g(x_r^A) = g(x_r^A)$ for $r = 0, \ldots, p - 1$ holds true.

Definition 5.3. *Let boxes $D = A \times \mathrm{i}C$, $A, B, C \subset \mathbb{R}$, and $p \in \mathbb{N}$, $p \geq 2$, be given. Moreover, let the mapping $z \colon D \to \mathbb{C}$,*

$$z(x) := \mathrm{e}^{-2\pi\mathrm{i}x \frac{\mathrm{diam}\, B}{(p-1)N}}, \tag{5.5}$$

and the Chebyshev nodes $z_j^{DB} := z(x_j^A)$ mapped on the arc $\Gamma_p^{DB} := z(D)$ be given. Under the admissibility condition in the sense of Inequality (4.5) for A and B, the nodes are distinct and we extend the Lagrange polynomials in Equation (4.7) to $\widetilde{L}_k^{DB} \colon D \to \mathbb{C}$ and the interpolation operator (4.8) to $\mathcal{I}_p^{DB} \colon C(\Gamma_p^{DB}) \to \Pi_{p-1}(\mathbb{C})$,

$$\widetilde{L}_k^{DB}(z^{DB}) := \prod_{\substack{j=0 \\ j \neq k}}^{p-1} \frac{z - z_j^{DB}}{z_k^{DB} - z_j^{DB}}, \tag{5.6}$$

$$\mathcal{I}_p^{DB} \widetilde{g} := \sum_{j=0}^{p-1} \widetilde{g}(z_j^{DB}) \widetilde{L}_j^{DB}. \tag{5.7}$$

Theorem 5.4. *Let $p, N \in \mathbb{N}$, $p \geq 5$, $A, B \subset \mathbb{R}$ admissible be given. Moreover, let*

$$D_p := A \times \mathrm{i}\,\mathrm{diam}\, A \frac{p-1}{2\pi} \left[\log\left(1 - \frac{1}{p}\right), \log\left(1 + \frac{1}{p}\right) \right],$$

$g \in E_B(D_p)$, $g(x) := \sum_{k=1}^{M_2} \hat{g}_k \mathrm{e}^{2\pi\mathrm{i}\xi_k x/N}$, $\hat{g}_k \in \mathbb{C}$, $k = 1, \ldots, M_2$ be given. Then we have the error estimate

$$\|g - \mathcal{J}_p^{D_p B} g\|_{C(D_p)} \leq (1 + \|\tilde{\mathcal{I}}_p\|)\tilde{c}_p \|\hat{\mathbf{g}}\|_1,$$

where the constant $\tilde{c}_p$ is given by Equation (5.3) and $\tilde{\mathcal{I}}_p := \mathcal{I}_p^{\tilde{D}_p[-N/2, N/2]}$ with

$$\|\tilde{\mathcal{I}}_p\| = \sup_{\substack{h \in C(\tilde{\Gamma}_p) \\ \|h\|_{C(\tilde{\Gamma}_p)=1}}} \|\tilde{\mathcal{I}}_p h\|_{C(\tilde{\Gamma}_p)} \geq \|\mathcal{I}_p\| \geq 1.$$

Proof. The proof is analogue to the proof of Theorem 4.8. We dilate and shift the interval $D_p \mapsto \tilde{D}_p$ and $B \mapsto [-N/2, N/2]$ via the maps (4.17) and (4.22), respectively. Let $\tilde{g} \colon \tilde{D}_p \to \mathbb{C}$ defined as in (4.23). We have

$$\mathcal{J}_p^{D_p B} g(x) = \mathrm{e}^{2\pi \mathrm{i} c^B x / N} \mathcal{J}_p \tilde{g}(y), \quad \mathcal{J}_p := \mathcal{J}_p^{\tilde{D}_p[-N/2, N/2]}.$$

With the same arguments and the function $h \colon \tilde{\Gamma}_p \to \mathbb{C}$, as defined in (4.24), we obtain

$$\mathcal{J}_p \tilde{g}(y) = z^{-\frac{p-1}{2}} \tilde{\mathcal{I}}_p h(z).$$

Moreover, the interpolation obeys

$$\max_{x \in D_p} |g(x) - \mathcal{J}_p^{D_p B} g(x)| = \max_{y \in \tilde{D}_p} |\tilde{g}(y) - \mathcal{J}_p \tilde{g}(y)| = \max_{z \in \tilde{\Gamma}_p} \left| z^{-\frac{p-1}{2}} \left(h(z) - \tilde{\mathcal{I}}_p h(z) \right) \right|.$$

Since $z \in \tilde{\Gamma}_p$ implies $1 - \frac{1}{p} \le |z| \le 1 + \frac{1}{p}$, we conclude $\frac{1}{|z|} \le 1 - \frac{1}{p} \le 1$ and finally

$$\max_{x \in D_p} |g(x) - \mathcal{J}_p^{D_p B} g(x)| \le \max_{z \in \tilde{\Gamma}_p} \left| h(z) - \tilde{\mathcal{I}}_p h(z) \right|.$$

Using the truncated Taylor expansion $h_p \colon \tilde{\Gamma}_p \to \mathbb{C}$ of h, cf. (4.25), and applying Lemma 5.1, we conclude the assertion

$$\begin{aligned}
\max_{z \in \tilde{\Gamma}_p} |h(z) - \tilde{\mathcal{I}}_p h(z)| &\le \max_{z \in \tilde{\Gamma}_p} |h(z) - h_p(z)| + \max_{z \in \tilde{\Gamma}_p} \left| h_p(z) - \tilde{\mathcal{I}}_p h(z) \right| \\
&\le (1 + \|\tilde{\mathcal{I}}_p\|) \max_{z \in \tilde{\Gamma}_p} |h(z) - h_p(z)| \\
&\le (1 + \|\tilde{\mathcal{I}}_p\|) \tilde{c}_p \|\hat{\mathbf{g}}\|_1.
\end{aligned}$$

The inequality

$$\sup_{\substack{h \in C(\tilde{\Gamma}_p) \\ \|h\|_{C(\tilde{\Gamma}_p)} = 1}} \|\tilde{\mathcal{I}}_p h\|_{C(\tilde{\Gamma}_p)} \ge \|\mathcal{I}_p\|$$

follows from Equation (4.26) by an appropriate constant continuation of the function h to $\tilde{\Gamma}_p$. Moreover, $\mathcal{I}_p$ is a polynomial interpolation operator and reproduces polynomials up to a degree $p - 1$. Let $h \colon \Gamma_p \to \mathbb{C}$, $\|h\|_{C(\Gamma_p)} = 1$, the assertion follows since

$$\|\mathcal{I}_p\| \ge \|\mathcal{I}_p h\|_{C(\Gamma_p)} = \|h\|_{C(\Gamma_p)} = 1.$$

$\square$

Remark 5.5. *One aspect, why we do not pursue that idea is that we can only compute the sum (5.1) for $w_j \in X$, where $X = [0, N] \times \mathrm{i} N \frac{p-1}{2\pi} \left[\log(1 - \frac{1}{p}), \log(1 + \frac{1}{p}) \right]$ is a small strip around the real line depending on the approximation rank p. Let the nonharmonic bandwidth $N = 2^L$, $L = 10$, the level $\ell = 5$ and boxes*

$$D = [0, 2^\ell] \times \mathrm{i} 2^\ell \frac{2\pi}{p-1} \left[\log(1 - \frac{1}{p}), \log(1 + \frac{1}{p}) \right] \quad and \quad B = [0, 2^\ell],$$

be given. Draw nodes $a_j \in [0, 2^\ell]$ and $\xi_j \in B$, $k = 1, \dots, 1024$, at random from the uniform distribution. Moreover, we choose equispaced points $b_k \in \left[\log(1 - \frac{1}{p}), \log(1 + \frac{1}{p})\right]$, $k = 1, \dots, 20$, and set the spatial nodes

$$\widetilde{X} := \left\{ a_j + \mathrm{i} 2^\ell \frac{2\pi}{p-1} b_k : j = 1, \dots, 1024, \ k = 1, \dots, 20 \right\}.$$

The relative error $\|\mathbf{F} - \widetilde{\mathbf{F}}\|_{1 \to \infty}$ is illustrated in Figure 5.2(a), where $\mathbf{F}$ denotes the exact matrix and $\widetilde{\mathbf{F}}$ the approximate matrix as described in Section 4.5.1. This figure illustrates that the error decreases up to an approximation rank $p = 15$, where the relative error is 10^{-5}. This slowly decreasing behavior for $p \leq 15$ and increasing effect for $p \geq 16$ might be due to the behavior of the interpolation operator norm $\|\tilde{\mathcal{I}}_p\|$. Let $g \in C(\tilde{\Gamma}_p)$ with $g(z_0) = 1$ and $g(z_j) = 0$ for $j = 1, \dots, p-1$ and $\|g\|_{C(\tilde{\Gamma}_p)} = 1$, then the norm of the interpolation operator obeys $\|\tilde{\mathcal{I}}_p\| \geq \|\tilde{\mathcal{I}}_p g\|_{C(\tilde{\Gamma}_p)} = \max_{z \in \tilde{\Gamma}_p} |\tilde{L}_0^{\tilde{D}_p[-N/2, N/2]}(z)|$. This maximal value is illustrated in Figure 5.2(b) in dependence on p, where the line represents the maximal value and the dashed line represents the increasing behavior $C \cdot 1.1397^p$ with some constant $C > 0$. Note, that Theorem 5.4 does not give an upper bound on the interpolation operator norm and thus we have no guarantee for a small local error.

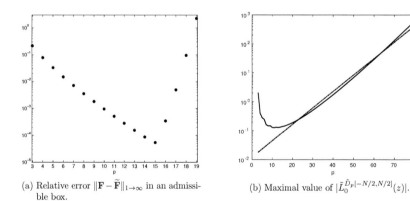

(a) Relative error $\|\mathbf{F} - \widetilde{\mathbf{F}}\|_{1 \to \infty}$ in an admissible box.

(b) Maximal value of $|\tilde{L}_0^{\tilde{D}_p[-N/2, N/2]}(z)|$.

Figure 5.2: Relative Error and maximal value of $|\tilde{L}_0^{\tilde{D}_p[-N/2, N/2]}(z)|$ for increasing approximation rank p.

5.2 Combining Laplace and Fourier transforms

Another idea to compute the sum (5.1) is a pointwise combination of Fourier transforms, i.e. the FFT, NFFT, NNFFT, or the BSFFT, and the Laplace transform from Chapter 3. Let the nonharmonic bandwidth $N = 2^L$, $L \in \mathbb{N}$, the domains $X, \Omega = [0, N]$ and sampling sets $\widetilde{X}, \widetilde{\Omega}$ as in the butterfly setting, see the beginning of Chapter 4, be given. Furthermore, let $\widetilde{Y} := \{y_j \in \mathbb{R} : j = 1, \ldots, M_1, \quad 0 < y_{M_1} < \cdots < y_1\}$. We denote the complex spatial nodes by $\widetilde{Z} := \{w_j \in \mathbb{C} : w_j = x_j + \mathrm{i}\frac{N}{2\pi} y_j : j = 1, \ldots, M_1\}$. We present a method to compute (5.1) for $\xi_k \in \widetilde{\Omega}$, $k = 1, \ldots, M_2$, and $w_j \in \widetilde{Z}$, $j = 1, \ldots, M_1$. We write

$$\mathrm{e}^{2\pi \mathrm{i} w_j / N} = \mathrm{e}^{2\pi \mathrm{i} x_j / N} \mathrm{e}^{-y_j}$$

and note that the summation (5.1) is a matrix vector multiplication with the matrix

$$\mathbf{C} := \mathbf{F} \odot \mathbf{K},$$

where $\mathbf{F}$ denotes the Fourier matrix given by (4.2) and $\mathbf{K}$ the Laplace matrix (3.2) with $\kappa(y, \xi) = \mathrm{e}^{-y\xi}$. We have the following result when approximating the factors as in the previous sections.

Lemma 5.6. *Let $\varepsilon \in (0, 1)$ and the Fourier and the Laplace matrix be approximated by*

$$\|\mathbf{F} - \tilde{\mathbf{F}}\|_{1 \to \infty} \leq \frac{\varepsilon}{3}, \quad \|\mathbf{K} - \tilde{\mathbf{K}}\|_{1 \to \infty} \leq \frac{\varepsilon}{3},$$

then

$$\|\mathbf{F} \odot \mathbf{K} - \tilde{\mathbf{F}} \odot \tilde{\mathbf{K}}\|_{1 \to \infty} \leq \varepsilon.$$

Proof. The estimate simply follows from

$$\|\mathbf{F}\|_{1 \to \infty} = 1, \quad \|\tilde{\mathbf{K}}\|_{1 \to \infty} = \|\tilde{\mathbf{K}} - \mathbf{K}\|_{1 \to \infty} + \|\mathbf{K}\|_{1 \to \infty} \leq 1 + \frac{\varepsilon}{3},$$

and

$$\mathbf{F} \odot \mathbf{K} - \tilde{\mathbf{F}} \odot \tilde{\mathbf{K}} = \mathbf{F} \odot (\mathbf{K} - \tilde{\mathbf{K}}) + (\mathbf{F} - \tilde{\mathbf{F}}) \odot \tilde{\mathbf{K}}.$$

Applying the norm $\|\cdot\|_{1 \to \infty}$, the triangle inequality and the rule

$$\|\mathbf{A} \odot \mathbf{B}\|_{1 \to \infty} = \max_{i,j} |\mathbf{A}_{i,j} \mathbf{B}_{i,j}| \leq \max_{i,j} |\mathbf{A}_{i,j}| \max_{i,j} |\mathbf{B}_{i,j}| = \|\mathbf{A}\|_{1 \to \infty} \|\mathbf{B}\|_{1 \to \infty}$$

for matrices $\mathbf{A}, \mathbf{B} \in \mathbb{C}^{M_1 \times M_2}$ twice yields the assertion.

$\square$

Note that $\mathbf{F}$ in Lemma 5.6 could be any Fourier matrix, for example the Fourier matrix of the FFT, NFFT or NNFFT approach.

We use the hierarchical decomposition of the discrete Laplace transform and realize matrix vector products with matrix blocks by the following technique.

Lemma 5.7. *By slight abuse of notation, let* $\mathbf{K} = (k_{i,j})_{i,j=1}^{m_1,m_2} = \mathbf{L}^Y \mathbf{K}^{Y,\Omega} \left(\mathbf{L}^\Omega\right)^\top$ *denote a single matrix block of the Laplace matrix, where* $\mathbf{L}^Y \in \mathbb{R}^{m_1 \times q}$, $\mathbf{K}^{Y,\Omega} \in \mathbb{R}^{q \times q}$, *and* $\mathbf{L}^\Omega \in \mathbb{R}^{m_2 \times q}$, *and* $\mathbf{F} = (f_{i,j})_{i,j=1}^{m_1,m_2}$ *the associated block of the Fourier matrix, then*

$$(\mathbf{F} \odot \mathbf{K})\hat{\mathbf{f}} = \left(\mathbf{L}^Y \odot \left(\mathbf{F} \operatorname{diag}(\hat{\mathbf{f}})\mathbf{L}^\Omega \left(\mathbf{K}^{Y,\Omega}\right)^\top\right)\right)\mathbf{1},$$

where $\mathbf{1} := (1, \ldots, 1)^\top \in \mathbb{R}^q$.

Proof. We define vectors $\mathbf{l}^Y := (l_i^Y)_{i=1,\ldots,m_1}$ and $\mathbf{l}^\Omega := (l_j^\Omega)_{j=1,\ldots,m_2}$. The simplest case is $\mathbf{K} = \mathbf{l}^Y \cdot 1 \cdot \left(\mathbf{l}^\Omega\right)^\top$. Hence $(\mathbf{F} \odot \mathbf{K})_{i,j} = f_{i,j}k_{i,j} = l_i^Y f_{i,j} l_j^\Omega$ implies

$$((\mathbf{F} \odot \mathbf{K})\hat{\mathbf{f}})_i = l_i^Y \sum_{j=1}^{m_2} f_{ij} l_j^\Omega \hat{f}_j = l_i^Y \sum_{j=1}^{m_2} f_{i,j}(\operatorname{diag}(\hat{\mathbf{f}})\mathbf{l}^\Omega)_j$$

and finally

$$(\mathbf{F} \odot \mathbf{K})\hat{\mathbf{f}} = \left(\mathbf{l}^Y \odot \left(\mathbf{A} \operatorname{diag}(\hat{\mathbf{f}})\mathbf{l}^\Omega \left(\mathbf{K}^{Y,\Omega}\right)^\top\right)\right) \cdot 1$$

This implies the result since

$$\mathbf{K} = \mathbf{L}^Y \mathbf{K}^{Y,\Omega} \left(\mathbf{L}^\Omega\right)^\top = \sum_{r=1}^q \mathbf{l}_r^Y \left(\tilde{\mathbf{l}}_r^{Y,\Omega}\right)^\top,$$

where $\mathbf{l}_r^Y$ and $\tilde{\mathbf{l}}_r^{Y,\Omega}$ denote the columns of $\mathbf{L}^Y$ and $\mathbf{L}^\Omega \left(\mathbf{K}^{Y,\Omega}\right)^\top$, respectively. $\square$

Theorem 5.8. *Let* $N, M_1, M_2 \in \mathbb{N}$ *with* $M_1, M_2 = \mathcal{O}(N)$, $\varepsilon > 0$, $w_j \in \widetilde{Z}$ *for* $j = 1, \ldots, M_1$, $\hat{\mathbf{f}} \in \mathbb{C}_2^M$, *and* $\mathbf{f} := \mathbf{C}\hat{\mathbf{f}}$ *with*

$$\mathbf{C} = \left(e^{2\pi i w_j \xi_k/N}\right)_{j,k=1}^{M_1,M_2} = \left(e^{2\pi i x_j \xi_k/N} e^{-y_j \xi_k}\right)_{j,k=1}^{M_1,M_2}$$

be given. Algorithm 7 takes

$$\mathcal{O}\left(N \log N \log \frac{N y_1}{\varepsilon} \log \frac{1}{\varepsilon}\right) \quad or \quad \mathcal{O}\left(N \log N \log^2 \frac{N}{\varepsilon} \log \frac{N y_1}{\varepsilon} \log \frac{1}{\varepsilon}\right)$$

floating point operations using the NNFFT or the BSFFT, respectively. Its output $\tilde{\mathbf{f}} \in \mathbb{C}^{M_1}$ *fulfills the error estimate*

$$\|\mathbf{f} - \tilde{\mathbf{f}}\|_\infty \leq \varepsilon \|\hat{\mathbf{f}}\|_1.$$

Proof. We write all nodes w in the form $w = x + i\frac{N}{2\pi}y \in \widetilde{Z}$ for the whole proof. We use the decomposition of the frequency nodes and the spatial nodes in dyadic intervals Ω_ℓ, Y_m, $\ell, m = 1, \ldots, M$, cf. Definition 3.7 of the Laplace transform. The decomposition of Y induces a decomposition of the nodes $\{x_j : j = 1, \ldots, N\} = \dot{\bigcup}_{m=1,\ldots,M} X_m$, where $X_m := \{x_j : y_j \in Y_m\}$. Accordingly, we denote restrictions of the Fourier matrix, the Fourier coefficients, and the result vector by superscripts with these sets. Due to Lemma 3.8 (i), we set

$$\tilde{\mathbf{f}}^{Y_M} := \tilde{\mathbf{F}}^{X_M,\Omega}\hat{\mathbf{f}}.$$

Let $y \in Y_m$, $m = 1, \ldots, M - 1$ and $\Psi \subset \Omega$. We denote by $f^\Psi \colon \mathbb{C} \to \mathbb{C}$ the function $f^\Psi(w) := \sum_{\xi_k \in \Psi} \hat{f}_k e^{2\pi i w \xi_k / N}$. Note that for $y \in Y_m$ the function f can be split in the sums

$$f(w) = f^{\Psi_1}(w) + f^{\Psi_2}(w) + f^{\Psi_3}(w),$$

where $\Psi_1 := \cup_{\ell < \ell_m} \Omega_\ell$, $\Psi_2 := \cup_{\ell = \ell_m}^{L_m} \Omega_\ell$, and $\Psi_3 := \cup_{\ell > L_m} \Omega_\ell$. Since $\xi_k \in \Psi_1$, implies $|f^{\Psi_1}(w)| < \varepsilon \|\hat{f}\|_1$ by applying Lemma 3.8 (ii), i.e., $f^{\Psi_1}(w_j)$ can be approximated by zero. Regarding the most interesting part of the approximation, the computation of $f^{\Psi_2}(w_j)$. The decomposition of $\tilde{K}$ in (3.7) yields

$$\mathbf{F} \odot \tilde{\mathbf{K}} = \left(\mathbf{F}^{X_m, \Omega_\ell} \odot \mathbf{L}^{Y_m} \mathbf{K}^{Y_m, \Omega_\ell} \left(\mathbf{L}^{\Omega_\ell} \right)^\top \right)_{m=1, \ell = \ell_m}^{M-1, L_m}.$$

For notational simplicity, we apply Lemma 5.7 to one block row m and two (artificial) block columns $\ell = 0, 1$, $\Psi_2 := \Omega_0 \cup \Omega_1$, in

$$\left(\mathbf{F}^{X_m, \Omega_0} \odot \mathbf{L}^{Y_m} \mathbf{K}^{Y_m, \Omega_0} \left(\mathbf{L}^{\Omega_0} \right)^\top \quad \mathbf{F}^{X_m, \Omega_1} \odot \mathbf{L}^{Y_m} \mathbf{K}^{Y_m, \Omega_1} \left(\mathbf{L}^{\Omega_1} \right)^\top \right) \hat{\mathbf{f}}^{\Psi_2}$$

$$= \left(\mathbf{L}^{Y_m} \odot \mathbf{F}^{X_m, \Omega_0} \left(\operatorname{diag} \hat{\mathbf{f}}^{\Omega_0} \right) \mathbf{L}^{\Omega_0} \left(\mathbf{K}^{Y_m, \Omega_0} \right)^\top \right) \mathbf{1}$$

$$+ \left(\mathbf{L}^{Y_m} \odot \mathbf{F}^{X_m, \Omega_1} \left(\operatorname{diag} \hat{\mathbf{f}}^{\Omega_1} \right) \mathbf{L}^{\Omega_1} \left(\mathbf{K}^{Y_m, \Omega_1} \right)^\top \right) \mathbf{1}$$

$$= \left(\mathbf{L}^{Y_m} \odot \mathbf{F}^{X_m, \Psi_2} \left(\begin{Bmatrix} \operatorname{diag} \hat{\mathbf{f}}^{\Omega_0} \\ \operatorname{diag} \hat{\mathbf{f}}^{\Omega_1} \end{Bmatrix} \begin{Bmatrix} \mathbf{L}^{\Omega_0} \left(\mathbf{K}^{Y_m, \Omega_0} \right)^\top \\ \mathbf{L}^{\Omega_1} \left(\mathbf{K}^{Y_m, \Omega_1} \right)^\top \end{Bmatrix} \right) \right) \mathbf{1}.$$

Now, the error estimate is a straightforward consequence of Lemma 5.6. The complexity estimate follows when considering the dominant computation in the second last line $\mathbf{H} = \ldots$ in Algorithm 7. We have $M = \mathcal{O}(\log \frac{N y_1}{\varepsilon})$ steps in the outer loop and $q = \mathcal{O}(\log \frac{1}{\varepsilon})$ right hand sides for the multiplication with the approximate Fourier matrix $\tilde{\mathbf{F}}$, whose computational needs are given below (4.64) for the BSFFT and $\mathcal{O}(N \log N)$ for the NNFFT, see [15]. $\qquad \square$

For abbreviation, we denote Algorithm 7 as XFLT (fast Fourier transform for complex evaluation nodes).

5.2.1 Evaluation of polynomials in the unit disk

An application for the computation of sums of the form (5.1) is the evaluation of generalized polynomials $f \colon \mathbb{C} \to \mathbb{C}$,

$$f(z) = \sum_{k=1}^{M_2} \hat{f}_k z^{\xi_k} \tag{5.8}$$

at nodes $z_j \in Z := \{ z \in \mathbb{C} : |z| \leq 1 \}$, $j = 1, \ldots, M_1$, and for exponents $\xi_k \in [1, N]$, $N \in \mathbb{N}$, where we exclude the nonpositive real axis $\operatorname{Re}(z_j) \leq 0$ for noninteger exponents ξ_k. We write

$$z = e^{-y} e^{2\pi i x / N}, \quad y \in [0, \infty), \ x \in [0, N). \tag{5.9}$$

Algorithm 7 Fast Fourier transform for complex evaluation nodes (XFLT).

Input:

$\varepsilon \in (0,1)$ $\triangleright$ target accuracy

$N \in \mathbb{N}$ $\triangleright$ number of sampling nodes

$\tilde{\Omega}$ $\triangleright$ nodes in frequency domain

$\tilde{Z}$ $\triangleright$ nodes in spatial domain

$\hat{\mathbf{f}} \in \mathbb{C}^{M_2}$ $\triangleright$ Fourier coefficients

Output:

$\tilde{\mathbf{f}} \in \mathbb{C}^{M_1}, \tilde{\mathbf{f}} \approx \mathbf{C}\hat{\mathbf{f}}$

$M = \left\lceil \log_2 \frac{N y_{M_1}}{\varepsilon} \right\rceil + 1$ $\triangleright$ number of decompositions

$q = \lceil \frac{1}{2} + \log_4 1/\varepsilon \rceil$ $\triangleright$ approximation rank, Laplace transform

$\tilde{\mathbf{f}} := 0$

$\tilde{\mathbf{f}}^{Y_M} = \bar{\mathbf{F}}^{Y_M, \Omega} \hat{\mathbf{f}}$

for $\ell = 1, \ldots, M-1$ **do**

$\quad \hat{\mathbf{F}}^{\Omega_\ell} := \left(\text{diag}\, \hat{\mathbf{f}}^{\Omega_\ell} \right) \mathbf{L}^{\Omega_\ell} \in \mathbb{C}^{|\Omega_\ell| \times q}$

end for

for $m = 1, \ldots, M-1$ **do**

$\quad \Psi := \cup_{\ell > L_m} \Omega_\ell$

$\quad \tilde{\mathbf{f}}^{Y_m} = \bar{\mathbf{F}}^{X_m, \Psi} \hat{\mathbf{f}}^\Psi$

$\quad \Psi := \cup_{\ell=\ell_m}^{L_m} \Omega_\ell$

$\quad \mathbf{H} := \bar{\mathbf{F}}^{X_m, \Psi} \left(\hat{\mathbf{F}}^{\Omega_\ell} \left(\mathbf{K}^{Y_m, \Omega_\ell} \right)^\top \right)_{\ell=\ell_m, \ldots, L_m} \in \mathbb{C}^{|X_m| \times q}$

$\quad \tilde{\mathbf{f}}^{Y_m} = \tilde{\mathbf{f}}^{Y_m} + \left(\mathbf{L}^{Y_m} \odot \mathbf{H} \right) \mathbf{1}$

end for

Due to $\xi_k \geq 1$ the condition $|z_j| \leq \varepsilon$ implies immediately $|f(z_j)| \leq \varepsilon$ and $y_j \leq \log \frac{1}{\varepsilon}$ for all $j = 1, \ldots, M_1$. We collect the associated nodes $Z^0 := \{z_j \in \mathbb{C} : |z_j| < \varepsilon\}$ and set the result $\tilde{\mathbf{f}} \in \mathbb{C}^N$, restricted to these nodes, to zero. Now write all nodes in polar form (5.9), where we can assume $0 \leq y_N \leq \ldots \leq y_1 \leq \log \frac{1}{\varepsilon}$. Furthermore, we denote by Y^0 and X^0 the associated nodes of the polar form of $z_j \in Z^0$. The only parts which change in Algorithm 7, are the parameter $M = \left\lceil \log_2 \frac{N \log 1/\varepsilon}{\varepsilon} \right\rceil + 1$ and that the appropriate nodes x_j and y_j of $z_j \in Z^0$ have to be excluded in the sets Y_m and X_m in the algorithm. In other words, we have to replace Y_m by $Y_m \setminus Y^0$ and X_m by $X_m \setminus X^0$ for all $m = 1, \ldots, M$.

Remark 5.9. *Note that [1] suggests a fast algorithm for the multiplication with $\boldsymbol{C}$ when the nodes z_j are close to the unit circle. Regarding generalizations, we get a fast algorithm for the multiplication with the adjoint matrix $\boldsymbol{C}^*$ simply using the adjoint* *algorithms for the matrices $\boldsymbol{K}$ and $\boldsymbol{A}$. In particular, this allows to evaluate $g : \mathbb{C} \to \mathbb{C}$,*

$$g(\xi) = \sum_{j=1}^{N} \hat{g}_j z_j^\xi$$

at nodes $\xi_k \in [1, N]$, $k = 1, \ldots, N$, and for given $z_j \in Z = \{z \in \mathbb{C} : |z| \leq 1\}$ and coefficients $\hat{g}_j \in \mathbb{C}$ efficiently. Possible applications include the fast evaluation of certain special functions when approximated as in [6] on the real line. The most general case with kernel

$$e^{(\xi + i\eta)(x + iy)} = e^{\xi x} e^{-by} e^{i(\eta x + \xi y)}$$

allows for efficient treatment when (ξ, η) as well as (x, y) sample a smooth contour in $\mathbb{C}$ and are in appropriate ranges. Then the last term leads to a 2d sparse FFT [58, 35] and we might apply the pointwise product idea twice.

5.3 Numerical experiments

We use the same hard- and software as described in Section 3.4. Again, we draw random uniformly distributed coefficients $\hat{f}_k \in [-\frac{1}{2}, \frac{1}{2}] \times [-\frac{1}{2}, \frac{1}{2}]i$, and sampling nodes $x_j \in [0, N]$ for $k = 1, \ldots, N$, and we draw random nodes $0 \leq y_N \leq y_{N-1} \leq \cdots \leq y_1 \leq (2q-1) \log 2$ which ensures $y_j \in [0, \log 1/\varepsilon]$ and $M = \lceil \log_2 \frac{N \log 1/\varepsilon}{\varepsilon} \rceil + 1$. Figure 5.3 compares the relative error ε_1 of Algorithm 7 for uniformly distributed frequencies $\xi_k \in [0, N]$, $k = 1, \ldots, N$, and for equispaced frequencies $\xi_k = 1, \ldots, N$, see Section 5.2.1. The error of the NFFT is given by

$$\|\mathbf{F} - \widetilde{\mathbf{F}}\|_{1 \to \infty} \leq \frac{4\pi}{2^{1/4}} e^{-\sqrt{2}\pi m} (\sqrt{m} + m),$$

where $m = \lceil q/3 \rceil$ denotes the approximation rank of the NFFT, see [33, page 11]. We use the default settings of the NFFT toolbox, where the Kaiser-Bessel functions are used as ansatz functions. The error of the NNFFT can also be bounded for this ansatz, cf. the error estimate [14, Satz 3.3] for the ansatz with Gaußkernels. For the combination of the Laplace transform with the BSFFT, we choose the approximation rank of the butterfly approach $p = \lceil \frac{q}{2} \rceil + 3$. Since $\|\mathbf{F}\|_{1 \to \infty} = 1$ and due to Lemmata 5.6 and 3.8, we obtain an error bound for an arbitrary Fourier transform

$$\|\mathbf{F} \odot \mathbf{K} - \tilde{\mathbf{F}} \odot \tilde{\mathbf{K}}\|_{1 \to \infty} \leq 2^{1-2q} + (1 + 2^{1-2q}) \|\mathbf{F} - \tilde{\mathbf{F}}\|_{1 \to \infty}. \tag{5.10}$$

The computational times of Algorithm 7 are shown in Figure 5.4. It illustrates the combination of the Laplace transform with the BSFFT and with the NFFT or NNFFT. In both cases all approximate algorithms have a complexity $\mathcal{O}(N \log N)$. But still, the constant of the BSFFT is very large.

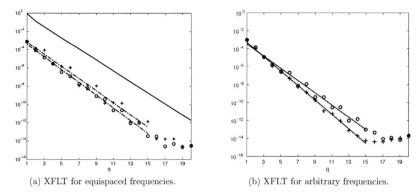

(a) XFLT for equispaced frequencies.   (b) XFLT for arbitrary frequencies.

Figure 5.3: Approximation error (3.9) with respect to the local expansion degree q and $N = 2^{14}$. The relative error of the BSFFT (circle), the NFFT (left, plus) or NNFFT (right, plus) ansatz. The dashed lines represent a least squares fit $\varepsilon \approx CK^{-2q}$ to the error and the solid line in the error bound (5.10) for the NFFT variant.

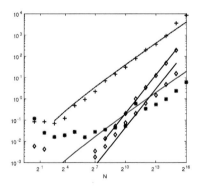

Figure 5.4: Behavior of the computational times for fixed approximation rank $q = 4$ with respect to the bandwidth N, the BSFFT (plus) with a least squares fit $CN \log N$ (dashed line), the naive computations (diamonds) with a least squares fit CN^2 (line) and the NFFT (asterisk) with a least squares fit $CN \log N$ (dotted line), respectively.

In our last experiments we compare [1, Algorithm 2], stated here as Algorithm 8, with Algorithm 7 for the NFFT case. Note that the presented algorithms in [1] allows for complex spatial and frequency nodes. These algorithms are based on a combination of convolutions and computing FFTs. Algorithm 8 computes for a bandwidth $N \in 2\mathbb{N}$, a fixed real parameter $A > 1$, the complex nodes $z_j = e^{a_j} e^{-2\pi i b_j}$, $j = 1, \ldots, M_1$, with $|a_j| \leq \frac{\log A}{N}$, $b_j \in [-1/2, 1/2]$, approximations $\widetilde{u}_j$ to the sums

$$u_j := \sum_{k=-N/2}^{N/2} \hat{u}_k z_j^k, \quad \text{for } j = 1, \ldots, M_1. \tag{5.11}$$

The error of this algorithm is bounded by $\|\mathbf{u} - \widetilde{\mathbf{u}}\|_\infty \leq \|\mathbf{u} - \widetilde{\mathbf{u}}\|_1 \leq \|\hat{\mathbf{u}}\|_1 (2\varepsilon + \mathrm{o}(\varepsilon))$ for the target accuracy ε, cf. [1, Theorem 2]. Note that the restriction on the nodes a_j implies a restriction to the absolute values of the complex nodes $A^{-1/N} \leq |z_j| \leq A^{1/N}$. This allows for the computation of the sums (5.11) via Algorithm 8 only in a concentric band with inner radius $A^{-1/N}$ and outer radius $A^{1/N}$. In other words, the sum (5.11) is equivalent to compute (5.1) with

$$w_j = b_j - i\frac{a_j}{2\pi} \in [-1/2, 1/2] \times \frac{1}{2\pi}[-\frac{\log A}{N}, \frac{\log A}{N}], \tag{5.12}$$

which describes only a small strip in the complex plane around the real line.
Defining $\hat{g}_k := \hat{f}_k e^{\pi i k}$ and applying some shifts to (5.1) in the settings of Algorithm 7 leads to

$$\sum_{k=1}^{N} \hat{f}_k e^{-y_j k} e^{2\pi i x_j k/N} = e^{-y_j(N/2+1)} e^{-2\pi i (\frac{1}{2} - \frac{x_j}{N})(\frac{N}{2}+1)} \sum_{k=-N/2}^{N/2-1} \hat{g}_{k+\frac{N}{2}+1} e^{-y_j k} e^{-2\pi i (\frac{1}{2} - \frac{x_j}{N})k}.$$

Since $x_j \in [0, N]$ yields $(\frac{1}{2} - \frac{x_j}{N}) \in [-1/2, 1/2]$ and $a_j := -y_j$ with the previous restriction and positive values $0 \leq y_j \leq \frac{\log A}{N}$, the last sum can be computed by Algorithm 8. This restriction on the real part of the transform (5.1) allows for the evaluation of sums $\sum_k \hat{f}_k z_j^k$ for $z_j \in \mathbb{C}$ in a small strip above the real line.
We use the oversampling parameter $\nu = 2$ in Algorithm 8 for all numerical experiments. In a first experiment, we compare the theoretical accuracy ε of the algorithms against the numerical error ε_1 for fixed bandwidth $N = 1024$ and a fixed parameter $y_1 := \log A/N$. We draw coefficients $\hat{f}_k \in [-\frac{1}{2}, \frac{1}{2}] \times [-\frac{1}{2}, \frac{1}{2}]i$ and source nodes $\xi_j \in [0, N]$, $y_j \in [0, \frac{\log A}{N}]$, $j = 1, \ldots, N$, at random from the uniform distribution. Figure 5.5 shows $\varepsilon \approx \varepsilon_1$ up to machine precision.
In a second experiment, we choose $N = 1024$ and $\varepsilon = 10^{-8}$. We compare the computational times of the naive evaluation (5.1), which should be constant for fixed N, and Algorithm 7 and Algorithm 8 with respect to the increasing width $A = 10^m$ and $m \in \{10k : k = 1, \ldots, 20\}$ of the real part. We draw coefficients $\hat{f}_k \in [-\frac{1}{2}, \frac{1}{2}] \times [-\frac{1}{2}, \frac{1}{2}]i$ and source nodes $\xi_j \in [0, N]$, $y_j \in [0, \frac{\log A}{N}]$, $j = 1, \ldots, N$, at random from the uniform distribution. Furthermore, the unequally Spaced Fast Laplace Transform (USFLT) is getting unstable, in the sense that we obtain an accuracy $\varepsilon_1 > 10^{-8}$ of Algorithm 8 for values $A > 10^{70}$ which implies $y_1 > 0.1574$, see Figure 5.5(b).

Algorithm 8 Unequally Spaced Fast Laplace Transform (USFLT) ([1, Algorithm 2]).

Input:

$\varepsilon \in (0,1)$	▷ target accuracy
$\nu \in \mathbb{N}, \nu \geq 2$	▷ oversampling parameter
$A > 1$	▷ maximal exponential growth
$N \in 2\mathbb{N}$	▷ bandwidth parameter
$\{b_j : j = 1, \ldots, M_1\} \in [-1/2, 1/2]^{M_1}$	▷ complex part of spatial nodes
$\{a_j : j = 1, \ldots, M_1\} \in \left[-\frac{\log A}{N}, \frac{\log A}{N}\right]^{M_1}$	▷ real part of spatial nodes
$\hat{\mathbf{u}} \in \mathbb{C}^{M_2}$	▷ Fourier coefficients

Output:

$\tilde{\mathbf{u}} := (\tilde{u}_j)_{j=1}^N \in \mathbb{C}^N$, $\tilde{\mathbf{u}} \approx \mathbf{C}\hat{\mathbf{u}}$ with $\mathbf{C} = \left(e^{a_j k - 2\pi i x_j k}\right)_{j=1, k=-N/2}^{M_1, N/2-1}$

$\mu = \frac{2\nu-1}{2\nu N^2(\nu-1)} \log A + \log(1/\varepsilon)\frac{1}{\nu N^2(\nu-1)}$

$\varphi_a(t) = \sqrt{\frac{\pi}{\mu}} e^{-\frac{\pi^2}{\mu}(t-i\frac{a}{2\pi})^2}$

$\hat{\varphi}_a(v) = e^{-\mu v^2 + av}$ ▷ modulated Gaussian

$T_\varepsilon = \frac{1}{\pi}\sqrt{\mu \log\frac{1}{\varepsilon} + \frac{\mu^2 N^2}{4} + \frac{(\log A)^2}{4N^2}} \sim \frac{1}{\pi}\frac{\log(1/\varepsilon)}{4N}$ ▷ threshold parameter

$\varphi_a^\varepsilon = \begin{cases} \varphi_a(t), & \text{if } |t| \leq T_\varepsilon \\ 0, & \text{otherwise.} \end{cases}$ ▷ thresholded modulated Gaussian

$M = \lceil \nu N T_\varepsilon \rceil$

for $\ell = -\frac{N}{2}, \ldots, \frac{N}{2} - 1$ **do**
   $\hat{g}_l := \frac{\hat{u}_l}{\hat{\varphi}_0(l)}$
end for

for $k = -\frac{\nu N}{2}, \ldots \frac{\nu N}{2} - 1$ **do**
   $G_k = \frac{1}{\nu N}\sum_{l=-N/2}^{N/2-1} \hat{g}_l e^{-2\pi i\frac{kl}{\nu N}}$ ▷ computed via FFT of length νN
end for

for $k = -\frac{\nu N}{2} - M, \ldots, \frac{\nu N}{2} + M - 1$ **do**

$$H_k = \begin{cases} G_{k+\nu N} & \text{if } k < -\frac{\nu N}{2} \\ G_{k-\nu N} & \text{if } k \geq \frac{\nu N}{2} \\ G_k & \text{otherwise} \end{cases}$$

end for

for $j = 1, \ldots, M_1$ **do**
   $\tilde{u}_j = \sum_{k=\lfloor \nu N b_j \rfloor - M}^{\lceil \nu N b_j \rceil + M} H_k \varphi_{a_j}^\varepsilon\left(\frac{k}{\nu N} - b_j\right)$
end for

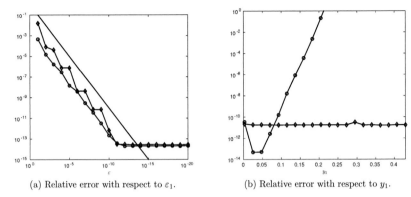

(a) Relative error with respect to ε_1.   (b) Relative error with respect to y_1.

Figure 5.5: The relative error ε_1 is illustrated for the NFFT variant (diamond) and the USFLT (circle) for $A = 1000$, $y_1 = \log A/N$, with respect to the target accuracy parameter ε (solid line) in Figure 5.5(a) and for the fixed target accuracy $\varepsilon = 10^{-8}$ with respect to $y_1 := \frac{\log A}{N}$ in Figure 5.5(b), both for a fixed bandwidth $N = 1024$.

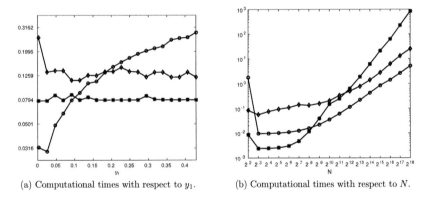

(a) Computational times with respect to y_1.   (b) Computational times with respect to N.

Figure 5.6: The behaviour ot the computational times for the NFFT variant (diamond), the USFLT (circle) and the naive computation (asterisk) is illustrated for a fixed bandwidth $N = 1024$ and $\varepsilon = 10^{-8}$ with respect to $y_1 := \frac{\log A}{N}$. Figure 5.6(a) and for a fixed parameter $A = 1000$, $y_1 := \frac{\log A}{N}$ with respect to the problem size N in Figure 5.6(b).

Due to the dependence of the summation parameter M on A, cf. Algorithm 8, the computational times of the USFLT increases for increasing values A whereas the NFFT variant is nearly constant as the naive exact computation, see Figure 5.6(a). For a fixed parameter $A = 1000$ and thus for a shrinking interval $Y = [0, y_1]$, $y_1 = \frac{\log A}{N}$, the USFLT and the NFFT approach have the same complexity rate $N \log N$, whereat the NFFT approach has a larger preconstant, see. Figure 5.6(b).

Chapter 6

Software library

The implementation of the algorithms presented in the previous chapters was an essential part of this thesis. All algorithms were implemented as a MATLAB toolbox called XFFT (eXtended fast Fourier transform). The toolbox consists of the butterfly transforms (Algorithm 3, 5), the Laplace transform (Algorithm 1), and the Fast Fourier transform for nonequispaced complex nodes (Algorithm 7). We give a short overview, how to use this toolbox. For readers who are not familiar with the class concept it is recommended to study the corresponding section of the MATLAB help, labeled as Object-Oriented Programming. The current release version XFFT 1.0 of the toolbox is available at http://sines.de.

under the GNU General Public License version 3 as published by the Free Software Foundation [18].

6.1 Butterfly sparse fast Fourier transform

The butterfly sparse Fourier transform is implemented for the dimensions $d = 1, 2, 3, 4$. The XFFT class consists of the main classes @sparse_FFT1D for $d = 1$, @sparse_FFT2D for $d = 2$, @sparse_FFT3D for $d = 3$, and @sparse_FFT4D for $d = 4$. To shorten notation, we write sparse_FFT*D, where the * can be chosen as 1,2,3,4. There are some classes called @tree1D, @tree2D, @tree3D, and @tree4D, which generate the trees of the dyadic decompositions of the domains X and Ω, see Algorithms 2 and 4. Note that the user must not call the tree-classes by oneself. The @sparse_FFT*D generates the tree automatically in a pre-computation step. The user has to set the properties N, MX, MOmega, p, and option given in Table 6.1. Here MX and MOmega are the spatial and frequency nodes X and Ω, respectively. Moreover, the user has to ensure that the input data fulfill the range conditions listed in Table 6.1, because there is no check for wrong input data implemented. If we want to choose a higher approximation rank $p \geq 6$ we should use for option the property 'Ltb' or 'Ltb*', because these variants are more stable, see Section 4.1.4 for details. Then we create an object of class @sparse_FFT*D,

$$\text{plan} = \text{sparse\_FFT} * D(\text{MX}, \text{MOmega}, \text{p}, \text{N}, \text{option}).$$

Afterwards we have to set the coefficient vector $\text{fhat} = (\hat{f}_k)_{k=1,\dots,M_2} \in \mathbb{C}^{M_2 \times 1}$ and can compute the sums $\mathbf{f} := (f(\boldsymbol{x}_j))_{j=1}^{M_1}$ in (4.1) approximatively by Algorithm 3 and 5, respectively, by applying

$$\mathbf{f} = \text{mtimes}(\text{plan}, \text{fhat}).$$

All available properties and methods of an object can be listed with the MATLAB functions properties and methods,

$$\text{properties}(\text{obj}) \quad \text{or} \quad \text{methods}(\text{obj}),$$

and documentation is provided by the MATLAB help and doc commands, for example

$$\text{help sparse\_FFT1D} \quad \text{or} \quad \text{doc sparse\_FFT1D}.$$

Property	Range	Description
N	$\text{N} = 2^L, L \in \mathbb{N}$	domain parameter
MX	$[0, N]^{M_1 \times d}$	sampling nodes $\widetilde{X}$
MOmega	$[0, N]^{M_2 \times d}$	sampling nodes $\widetilde{\Omega}$
p	$\text{p} \in \mathbb{N}$	local expansion degree
option	see Table 6.2	

Table 6.1: Properties of the class sparse_FFT*D for each dimension $d = 1, 2, 3, 4$ and their default values.

option	Description
'Mtb'	monomial-type basis, see Sections 4.1.3.1 and 4.2.3.1
'Ltb'	Lagrange-type basis, see Sections 4.1.3.2 and 4.2.3.2
'Ltb*'	Lagrange-type basis with pre-computation of the polynomials in MX.

Table 6.2: Possible types for the property option of the class sparse_FFT.

Remark 6.1. *In the implementation of the algorithms for $d = 2, 3, 4$ all matrices, which are defined as Kronecker products in Section 4.2.3 were defined as Kronecker products in reversed order. In this thesis we used for simplicity of notation for example $\mathbf{M}^{AB} = \bigotimes_{\nu=1}^{d} \mathbf{M}^{A_\nu B_\nu}$, see Section 4.2.3.1 for more details. We have with two appropriate permutation matrices $\mathbf{Q}, \mathbf{P} \in \{0, 1\}^{p^d \times p^d}$ the relation for the reversed ordering $\mathbf{P}\mathbf{M}^{AB}\mathbf{Q} = \bigotimes_{\nu=1}^{d} \mathbf{M}^{A_{d-\nu+1}B_{d-\nu+1}}$, see [47, page 26].*

6.2 Fast Fourier transform for complex evaluation nodes

The XFFT class consists of the main class @XFLT, which includes the fast Laplace transform, Algorithm 1, to compute

$$f_j := \sum_{k=1}^{M_2} \hat{f}_k e^{-y_j \xi_k}, \quad j = 1, \ldots, M_1. \tag{LT}$$

Furthermore, it includes Algorithm 7, the Fast Fourier transform for nonequispaced complex nodes, to compute sums of the form

$$f_j := \sum_{k=0}^{N-1} \hat{f}_k e^{2\pi i k j/N} e^{-y_j k}, \quad j = 0, \ldots, N-1, \tag{FFLT}$$

$$f_j := \sum_{k=1}^{N} \hat{f}_k e^{2\pi i k x_j/N} e^{-y_j k}, \quad j = 1, \ldots, M_1, \tag{NFLT}$$

$$f_j := \sum_{k=1}^{M_2} \hat{f}_k e^{2\pi i \xi_k x_j/N} e^{-y_j \xi_k}, \quad j = 1, \ldots, M_1,. \tag{(BSFLT) or (NNFLT)}$$

For more details, we refer the reader to the previous chapters. The first input parameter will set the form of the transform, (FLT), (NFLT), (BSFLT) or (NNFLT). The possible transforms are listed in Table 6.3.

transform, `Ftype`	Description
`'LT'`	Laplace transform
`'FFLT'`	fast Fourier Laplace transform
`'NFLT'`	nonequispaced fast Fourier Laplace transform
`'BSFLT'`	butterfly sparse fast Fourier Laplace transform
`'NNFLT'`	nonequispaced fast Fourier Laplace transform in spatial and frequency domain

Table 6.3: Possible transforms.

For each transform, the following values have to be set. First, the user has to set sampling nodes $\texttt{MY} = (y_1, \ldots, y_{M_1}) \in [0, \infty)^{M_1}$ in ascending order. Furthermore, the user can choose between the target accuracy ε or the approximation rank q of the Laplace transform. Finally the sampling nodes $\texttt{MOmega} = (\xi_1, \ldots, \xi_{M_2}) \in [0, \infty)^{M_2}$ has to be set in ascending order, too. For the Laplace transform, you can initialize the plan for fixed target accuracy `epsilon` by

```
plan=XFLT('LT',MY,'accuracy',epsilon,'MOmega',MOmega)
```

or for a fixed approximation rank q by

```
plan=XFLT('LT',MY,'rank',q,'MOmega',MOmega).
```

The computation of $\mathbf{f} = (f_j)_{j=1,\ldots,M_1}$ is done by the function call

    f=mtimes(plan,fhat),

where $\mathtt{fhat} = (\hat{f}_k)_{k=1,\ldots,M_2}$. For the NFLT and NNFLT the NFFT software library is needed, see [33] and

    http://www-user.tu-chemnitz.de/~potts/nfft/.

For an arbitrary plan, the 'Ftype' have to been choosen by possible values listed in Table 6.3. Moreover, the optional input arguments varargin are listed for each transformation in Table 6.4. The plan can be initialized by

    plan=XFLT(Ftype,MY,option,value,varargin),

where option has to be set as 'rank' or 'accuracy' with the appropriate value $\mathtt{q} \in \mathbb{N}$ or epsilon $\in (0,1)$, respectively.

Ftype Ftype	Property	Range	Description
LT	'MOmega'	$(\xi_1,\ldots,\xi_{M_2})^\top \in [0,N]^{M_2}$, $\xi_1 < \cdots < \xi_{M_2}$	frequency nodes
FFLT			everything is automatically set
NFLT	'N'	$N \in \mathbb{N}$	length of the NFFT
	'MX'	$(x_0,\ldots,x_{N-1})^\top \in [0,N]^N$	spatial nodes
	'libdir'	path	NFFT library directory
BSFLT	'N'	$N = 2^L, L \in \mathbb{N}$	domain parameter
	'MX'	$(x_1,\ldots,x_{M_1})^\top \in [0,N]^{M_1}$	spatial nodes
	'MOmega'	$(\xi_1,\ldots,\xi_{M_2})^\top \in [0,N]^{M_2}$,	sampling nodes in frequency domain
		$\xi_1 < \cdots < \xi_{M_2}$	
	'BSFFTrank'	$p \in \mathbb{N}$	local expansion degree
NFFLT	'N'	$N = 2^L, L \in \mathbb{N}$	domain parameter
	'MX'	$(x_1,\ldots,x_{M_1})^\top \in [0,N]^{M_1}$	spatial nodes
	'MOmega'	$(\xi_1,\ldots,\xi_{M_2})^\top \in [0,N]^{M_2}$, $\xi_1 < \cdots < \xi_{M_2}$	frequency nodes
	'libdir'	path	NFFT library directory

Table 6.4: Properties of the class XFLT and their default values.

7

Conclusion

In this thesis, we treated the computation of transforms with asymptotically smooth and oscillatory kernels. Recently, the butterfly approximation scheme and hierarchical approximations have been proposed for the efficient computation of integral transforms with oscillatory and with asymptotically smooth kernels.

In the first part of this thesis, we summarized and slightly improved the fast discrete Laplace transform [46] in a modern form including a generalization to asymptotically smooth kernels.

The second part of this thesis concerns the butterfly approximation scheme, which has been used for the development of a fast Fourier transform for sparse data [2, 58] which takes $\mathcal{O}(N^{d-1} \log N p^{d+1})$ floating point operations for $d \geq 2$, $M_1 = M_2 = \mathcal{O}(N^{d-1})$, well distributed sampling sets $\widetilde{\Omega}, \widetilde{X}$ on smooth $(d-1)$-dimensional manifolds, and a local expansion degree $p \in \mathbb{N}$. We presented a rigorous error analysis of this algorithm, showing that the local expansion degree grows at most as $p \approx |\log \varepsilon| + \log N$ and thus gave a complexity estimate for the scheme. Moreover, we showed theoretically as well as numerically that the original approach becomes numerically unstable if a large local expansion degree is used. We developed a stable variant by representing all approximations in a Lagrange type basis. One drawback of the butterfly sparse fast Fourier transform is a very huge pre-constant in the complexity. A further reduction in absolute computing time is necessary for real applications. A possible solution might be the parallelization of the butterfly scheme as done in [44].

Finally, we combined the Laplace transform with the Fourier transform in a purely algebraic fashion where we used the decomposition of the Laplace transform explicitly and a small number of generalized fast Fourier transforms as black box. This allows for a fast Fourier transform with nonequispaced nodes in the upper half plane. In this situation, the butterfly Fourier transform could be replaced by the nonequispaced fast Fourier transform which is both asymptotically as well as with respect to actual computation times faster. The Laplace transform as well as the fast Fourier transform for complex spatial nodes are presented in the univariate case but they can be generalized in a straightforward manner with tensor decompositions. The here presented algorithm is applicable for the evaluation of polynomials at many complex nodes in the unit disc, whereas the algorithm presented in [1] allows only the evaluation of complex nodes z in a small annulus $A^{-1/N} \leq |z| \leq A^{1/N}$, where N describes the polynomial degree and $A > 1$. Furthermore, our idea will also lead to a Fourier transform for samples $\widetilde{X}$ and $\widetilde{\Omega}$, both on a smooth contour in $\mathbb{C}$ with appropriate ranges for the spatial and frequency domain, see Remark 5.9. This issue is not treated in detail in this thesis and remains subject to future work.

Bibliography

[1]   F. Andersson. Algorithms for unequally spaced fast Laplace transforms. *Appl. Comput. Harmon. Anal.*, 35(3):419–432, 2013.

[2]   A. A. Aydıner, W. C. Chew, J. Song, and T. J. Cui. A sparse data fast Fourier transform (SDFFT). *IEEE Trans. Antennas and Propagation*, 51(11):3161–3170, 2003.

[3]   M. Bebendorf. Approximation of boundary element matrices. *Numer. Math.*, 86(4):565–589, 2000.

[4]   M. Bebendorf. *Hierarchical matrices*. Vol. 63 of *Lecture Notes in Computational Science and Engineering*. Springer-Verlag, Berlin, 2008, pp. xvi+290. A means to efficiently solve elliptic boundary value problems.

[5]   G. Beylkin. On the fast Fourier transform of functions with singularities. *Appl. Comput. Harmon. Anal.*, 2(4):363–381, 1995.

[6]   G. Beylkin and L. Monzón. On approximation of functions by exponential sums. *Appl. Comput. Harmon. Anal.*, 19(1):17–48, 2005.

[7]   P. L. Butzer and R. J. Nessel. *Fourier analysis and approximation*. Academic Press, New York-London, 1971, pp. xvi+553. Volume 1: One-dimensional theory, Pure and Applied Mathematics, Vol. 40.

[8]   E. Candès, L. Demanet, and L. Ying. A fast butterfly algorithm for the computation of Fourier integral operators. *Multiscale Model. Simul.*, 7(4):1727–1750, 2009.

[9]   J. W. Cooley and J. W. Tukey. An algorithm for the machine calculation of complex Fourier series. *Math. Comp.*, 19:297–301, 1965.

[10] L. Demanet, M. Ferrara, N. Maxwell, J. Poulson, and L. Ying. A butterfly algorithm for synthetic aperture radar imaging. *SIAM J. Imaging Sci.*, 5(1):203–243, 2012.

[11] A. Dutt and V. Rokhlin. Fast Fourier transforms for nonequispaced data. II. *Appl. Comput. Harmon. Anal.*, 2(1):85–100, 1995.

[12] A. Dutt and V. Rokhlin. Fast Fourier transforms for nonequispaced data. *SIAM J. Sci. Comput.*, 14(6):1368–1393, 1993.

[13] A. Edelman, P. McCorquodale, and S. Toledo. The future fast Fourier transform? *SIAM J. Sci. Comput.*, 20(3):1094–1114, 1999.

[14]   B. A. Elbel. Mehrdimensionale schnelle Fouriertransformation für nichtäquidistante Daten. Diplomarbeit. Technische Universität Darmstadt, 1998.

[15]   B. Elbel and G. Steidl. Fast Fourier transforms for nonequispaced data. In *Approximation theory IX, Vol. 2*. In Innov. Appl. Math. Vanderbilt Univ. Press, Nashville, TN, 1998, pp. 39–46.

[16]   T. Finck, G. Heinig, and K. Rost. An inversion formula and fast algorithms for Cauchy-Vandermonde matrices. *Linear Algebra Appl.*, 183:179–191, 1993.

[17]   L. Fox and I. B. Parker. *Chebyshev polynomials in numerical analysis*. Oxford University Press, London-New York-Toronto, Ont., 1968, pp. ix+205.

[18]   Free Software Foundation. GNU General Public License version 3 (GPLv3). http://www.gnu.org/licenses/. June 2007.

[19]   M. Frigo and S. G. Johnson. Pruned FFTs. http://www.fftw.org/pruned.html.

[20]   E.-M. Gassner. Approximation trigonometrischer Polynome auf einem Streifen. Bachelorarbeit. Universität Osnabrück, 2012.

[21]   C. F. Gauß. Nachlass: Theoria interpolationis methodo nova tractata. In, *Carl Friedrich Gauß Werke, Band III*. Vol. 3, pp. 265–327. Königliche Gesellschaft der Wissenschaften, Göttingen, 1866.

[22]   S. A. Goreinov and E. E. Tyrtyshnikov. The maximal-volume concept in approximation by low-rank matrices. Contemp. Math. 280:47–51, 2001.

[23]   T. Görner. Spherical mean values - Efficient computation by Fourier techniques and regularized reconstructions of function samples from discrete means. Dissertation, repOSitorium der Universität Osnabrück. July 2015.

[24]   T. Görner, R. Hielscher, and S. Kunis. Efficient and accurate computation of spherical mean values at scattered center points. *Inverse Probl. Imag.*, 6(4):645–661, 2012.

[25]   L. Grasedyck and W. Hackbusch. Construction and arithmetics of $\mathcal{H}$-matrices. *Computing*, 70(4):295–334, 2003.

[26]   L. Greengard and V. Rokhlin. A fast algorithm for particle simulations. *J. Comput. Phys.*, 73(2):325–348, 1987.

[27]   L. Greengard and J.-Y. Lee. Accelerating the nonuniform fast Fourier transform. *SIAM Rev.*, 46(3):443–454, 2004.

[28]   W. Hackbusch. A sparse matrix arithmetic based on $\mathcal{H}$-matrices. I. Introduction to $\mathcal{H}$-matrices. *Computing*, 62(2):89–108, 1999.

[29]   W. Hackbusch. *Hierarchische Matrizen: Algorithmen und Analysis*. Springer-Verlag, Berlin Heidelberg, 2009, pp. XX+451.

[30]   N. Hale and A. Townsend. A fast, simple, and stable Chebyshev-Legendre transform using an asymptotic formula. *SIAM J. Sci. Comput.*, 36(1):A148–A167, 2014.

[31] M. Hanke-Bourgeois. *Grundlagen der numerischen Mathematik und des wissenschaftlichen Rechnens.* Vieweg + Teubner, Wiesbaden, third ed., 2009, pp. 1–840.

[32] L. Kämmerer, S. Kunis, I. Melzer, D. Potts, and T. Volkmer. Computational methods for the Fourier analysis of sparse high-dimensional functions. In, *Extraction of quantifiable information from complex systems.* Vol. 102, in Lect. Notes Comput. Sci. Eng., pp. 347–363. Springer, Cham, 2014.

[33] J. Keiner, S. Kunis, and D. Potts. Using NFFT 3—a software library for various nonequispaced fast Fourier transforms. *ACM Trans. Math. Software,* 36(4):Art. 19, 30, 2009.

[34] T. G. Kolda and B. W. Bader. Tensor decompositions and applications. *SIAM Rev.,* 51(3):455–500, 2009.

[35] S. Kunis and I. Melzer. A stable and accurate butterfly sparse Fourier transform. *SIAM J. Numer. Anal.,* 50(3):1777–1800, 2012.

[36] S. Kunis and I. Melzer. Fast evaluation of real and complex exponential sums. *Preprint,* 2014.

[37] S. Kunis and I. Melzer. XFFT, Matlab toolbox for computing extended fast Fourier transforms: butterfly sparse Fourier transform, Laplace transform, fast Fourier transform for complex evaluation nodes. http://sines.de. 2015.

[38] H. J. Landau and H. O. Pollak. Prolate spheroidal wave functions, Fourier analysis and uncertainty. III. The dimension of the space of essentially time- and band-limited signals. *Bell System Tech. J.,* 41:1295–1336, 1962.

[39] R. R. Lederman. On the analytical and numerical properties of the truncated Laplace transform. Dissertation, Technical Report YALEU/DCS/TR-1490. May 2014.

[40] J. C. Mason and D. C. Handscomb. *Chebyshev polynomials.* Chapman & Hall/CRC, Boca Raton, FL, 2003, pp. xiv+341.

[41] I. Melzer. Schnelle Fourier-Transformation für dünne Daten. Diplomarbeit. Technische Universität Chemnitz, 2010.

[42] E. Michielssen and A. Boag. A multilevel matrix decomposition algorithm for analyzing scattering from large structures. *IEEE Trans. Antennas and Propagation,* 44(8):1086–1093, 1996.

[43] M. O'Neil, F. Woolfe, and V. Rokhlin. An algorithm for the rapid evaluation of special function transforms. *Appl. Comput. Harmon. Anal.,* 28(2):203–226, 2010.

[44] J. Poulson, L. Demanet, N. Maxwell, and L. Ying. A parallel butterfly algorithm. *SIAM J. Sci. Comput.,* 36(1):C49–C65, 2014.

[45] T. J. Rivlin. *The Chebyshev polynomials.* Wiley-Interscience [John Wiley & Sons], New York-London-Sydney, 1974, pp. vi+186. Pure and Applied Mathematics.

[46] V. Rokhlin. A fast algorithm for the discrete Laplace transformation. *J. Complexity,* 4(1):12–32, 1988.

[47] W.-H. Steeb. *Kronecker product of matrices and applications.* Bibliographisches Institut, Mannheim, 1991, pp. ii+139.

[48] G. Steidl. A note on fast Fourier transforms for nonequispaced grids. *Adv. Comput. Math.*, 9(3–4):337–352, 1998.

[49] J. Strain. A fast Laplace transform based on Laguerre functions. *Math. Comp.*, 58(197):275–283, 1992.

[50] X. Sun and N. P. Pitsianis. A matrix version of the fast multipole method. *SIAM Rev.*, 43(2):289–300 (electronic), 2001.

[51] A. Townsend. A Fast Analysis-Based Discrete Hankel Transform Using Asymptotic Expansions. *SIAM J. Numer. Anal.*, 53(4):1897–1917, 2015.

[52] L. N. Trefethen and J. A. C. Weideman. Two results on polynomial interpolation in equally spaced points. *J. Approx. Theory*, 65(3):247–260, 1991.

[53] L. N. Trefethen. *Approximation theory and approximation practice.* Society for Industrial and Applied Mathematics (SIAM), Philadelphia, PA, 2013, pp. viii+305.

[54] M. Tygert. Fast algorithms for spherical harmonic expansions, III. *J. Comput. Phys.*, 229(18):6181–6192, 2010.

[55] E. Tyrtyshnikov. Incomplete cross approximation in the mosaic-skeleton method. *Computing*, 64(4):367–380, 2000.

[56] G. N. Watson. *A Treatise on the Theory of Bessel Functions.* Cambridge University Press, Cambridge, GB, 1966, pp. vi+804.

[57] H. Widom. Asymptotic behavior of the eigenvalues of certain integral equations. II. *Arch. Rational Mech. Anal.*, 17:215–229, 1964.

[58] L. Ying. Sparse Fourier transform via butterfly algorithm. *SIAM J. Sci. Comput.*, 31(3):1678–1694, 2009.

[59] L. Ying, G. Biros, and D. Zorin. A kernel-independent adaptive fast multipole algorithm in two and three dimensions. *J. Comput. Phys.*, 196(2):591–626, 2004.

[60] L. Ying and S. Fomel. Fast computation of partial Fourier transforms. *Multiscale Model. Simul.*, 8(1):110–124, 2009.

[61] R. M. Young. *An introduction to nonharmonic Fourier series.* Academic Press Inc., San Diego, CA, first ed., 2001, pp. xiv+234.

Nomenclature

Abbreviations

BSFFT Butterfly sparse fast Fourier transform

FFT Fast Fourier transform

NFFT Nonequispaced fast Fourier transform

NNFFT Nonequispaced fast Fourier transform in space and frequency domain

USFLT Unequally spaced fast Laplace transform

XFFT Extended Fast Fourier transform

XFLT Fast Fourier transform for complex spatial nodes

Symbols

$\mathbf{A}^\top$ Transposed Matrix, page 8

$\mathbf{A}_k$ k-mode matricization of a tensor $\mathcal{A}$, page 11

$\mathbb{C}$ Complex numbers, page 7

$C(D)$ Space of continuous functions on D, page 7

c^A center of the box (interval) A, page 8

$C^{(m)}(D)$ Space of m-times continuously differentiable functions, page 7

C_p $C_p = K_p \left(1 + \frac{2}{\pi} \log p\right)$, page 37

c_p $c_p = \frac{1}{\pi p} \left(\frac{\pi}{p-1}\right)^p$, page 40

$\delta_{k,j}$ Kronecker delta, page 8

diag $\mathbf{v}$ Diagonal matrix with diagonal $\mathbf{v}$, page 8

diam Diameter, page 8

dist   Distance, page 8

∂_x Partial derivative, page 7

ε_1 Relative error of the approximation vector, page 30

$E_B(A)$ Space of all finite expansions of exponential functions $g\colon A \to \mathbb{C}$ with frequencies in B for dimension $d = 1$, page 35

$E_B^p(A)$ Space of all finite expansions of exponential functions $g\colon A \to \mathbb{C}$ with p equispaced frequencies in B for dimension $d = 1$, page 35

$E_B(A)$ Space of all finite expansions of exponential functions $g\colon A \to \mathbb{C}$ with frequencies in B for all dimensions, page 57

$E_B^p(A)$ Space of all finite expansions of exponential functions $g\colon A \to \mathbb{C}$ with p equispaced frequencies in B for all dimensions d, page 57

$\hat{f}_k, \hat{\mathbf{f}}$ k-th coefficient, coefficient vector

$\mathbf{f}, \widetilde{\mathbf{f}}$ Function values $\mathbf{f}$ and the corresponding approximation $\widetilde{\mathbf{f}}$ by the Laplace- or butterfly sparse Fourier transform

$f, \widetilde{f}$ Function f and the approximation $\widetilde{f}$ by the Laplace- or butterfly sparse Fourier transform

$f^{(m)}$ m-th derivative of f, page 8

Γ_p Sector of complex arc, page 40

Γ_p^{AB} Sector of complex arc, page 35

$(\mathcal{I}_q^A \otimes \mathcal{I})\kappa(y,\xi), (\mathcal{I} \otimes \mathcal{I}_q^B)\kappa(y,\xi)$ Interpolation polynomial of degree $p-1$ in one variable of $\kappa\colon A \times B \to \mathbb{R}$ in Chebyshev nodes in A or B respectively, page 22

$\mathcal{I}_p^A f$ Interpolation polynomial of f in the Chebyshev nodes in A of degree $p-1$, page 15

$\mathcal{I}_p^{AB}$ Interpolation operator $\mathcal{I}_p^{AB}\colon C(\Gamma_p^{AB}) \to \Pi_{p-1}(\mathbb{C})$, page 35

$\mathcal{I}_q^{A \times B}\kappa(y,\xi)$ Interpolation polynomial of degree $p-1$ in two variables $\kappa\colon A \times B \to \mathbb{R}$ in Chebyshev nodes in A and B, page 22

$\mathbf{I}_k$ Identity matrix of dimension $k \times k$, page 12

$\mathcal{J}_p$ Interpolation operator $\mathcal{J}_p^{[-\frac{1}{2}, \frac{1}{2}][-\frac{N}{2}, \frac{N}{2}]}$, page 42

$\mathcal{J}_p^{AB}$ Trigonometric interpolation operator, page 35

$\kappa(\mathbf{A})$ Condition number in the spectral norm of the matrix $\mathbf{A}$

$\kappa(x,\xi)$ Kernel function $\kappa\colon X \times \Omega \to \mathbb{R}$

$\mathbf{K}^{Y_m,\Omega_\ell}$ Matrix with entries of the evaluated kernel function in nodes Y_m and Ω_ℓ, page 23

K_p $\quad K_p = \left(\frac{2\pi^2}{\left(1-\cos\frac{2\pi}{p-1}\right)(p-1)^2}\right)^{\frac{p-1}{2}}$, page 37

ℓ $\quad$ Level of the dyadic decomposition, page 46

ℓ^p $\quad$ Sequence space, page 8

ℓ_m, L_m Lower and upper summation indices in the Laplace transform, page 23

$\mathbf{L}^{Y_m}, \mathbf{L}^{\Omega_m}$ Lagrange matrices, page 23

$\widetilde{L}_k^{AB}$ $\quad$ Lagrange polynomials to the nodes z_j^{AB}, page 35

l_r^{AB} $\quad$ Trigonometric Lagrange function, page 36

L_s, L_s^A s-th Lagrange polynomial to the Chebyshev nodes in $[-1,1]$ or A, page 15

$L_{\boldsymbol{x},k}$ $\quad$ k-th Lagrange polynomial to the nodes $\boldsymbol{x}$, page 13

$\mathbb{N}$ $\quad$ Natural numbers, page 7

$\mathbb{N}_0$ $\quad$ Natural numbers with 0, page 7

N $\quad$ Bandwidth, page 33

$\mathcal{O}, o$ $\quad$ Landau notation, page 9

$\widetilde{\Omega}$ $\quad$ Sampling nodes in the frequency domain, page 33

$\Omega_{\ell,n}$ $\quad$ n-th box in the ℓ-th dyadic decomposition of Ω, page 46

Ω $\quad$ Domain or Frequency domain

Ω_m $\quad$ Geometrical decompositions of Ω in the Laplace transform, page 23

$\Pi_p(\mathbb{K}_1, \mathbb{K}_2)$ Space of all polynomials mapping from $\mathbb{K}_1$ to $\mathbb{K}_2$ of degree at most p, page 7

$\mathbb{R}$ $\quad$ Real numbers, page 7

$\mathbb{R}^+$ $\quad$ Positive real numbers with 0, page 7

rank $\quad$ Rank of a matrix, page 8

$\mathbf{a}^*$ $\quad$ conjugate transpose vector, page 8

$\mathbf{a}^\top$ $\quad$ transpose vector, page 8

$\mathcal{T}_X$ $\quad$ Tree of the dyadic decomposition of X, page 47

$t_j, \boldsymbol{t}$ $\quad$ j-th Chebyshev nodes or Chebyshev nodes as vector, page 14

$t_j^{\max}$ $\quad$ Local extrema nodes of Chebyshev polynomial, page 14

T_p	p-th Chebyshev polynomial, page 14
U_p	Chebyshev polynomial of the second kind, page 14
vec	Vectorization of a matrix, page 9
vec	Vectorization of a tensor, page 11
$w_{\boldsymbol{x}}$	Nodal polynomial to the nodes $\boldsymbol{x}$, page 13
$\widetilde{X}$	Sampling nodes in spatial domain, page 33
X	Spatial domain, page 33
$X_{\ell,m}$	m-th box in the ℓ-th dyadic decomposition of X, page 46
$\boldsymbol{x}^A$	Chebyshev nodes in the box A, page 14
$\boldsymbol{x}_{\mathbf{r}}^A$	Chebyshev tensor grid in A, page 57
ξ_j^B	Equispaced nodes in the box B, page 35
$\boldsymbol{\xi}_{\mathbf{s}}^B$	Equispaced tensor grid in B, page 57
Y	Domain in the Laplace transform, page 23
Y_m	Geometrical decompositions of Y in the Laplace transform, page 23
$\mathbb{Z}$	Integers, page 7
z_j^{AB}	Chebyshev nodes on the complex arc, page 35
$\binom{\alpha}{k}$	Generalized binomial coefficient, page 8
$\lfloor\cdot\rfloor$, $\lceil\cdot\rceil$	Rounding down and up to the next integer, pointwise for vectors, page 9
$\|\cdot\|_p$, $\|\cdot\|_\infty$	Vector norms, page 8
$\|\cdot\|_{1\to\infty}$	Maximal absolute matrix entry, page 8
$\|\cdot\|_{C(D)}$	Maximum or supremum norm on D, page 7
$\|\cdot\|_2$	Spectral norm of a matrix, page 8
$\|\cdot\|_{\mathrm{F}}$	Frobenius norm of a matrix, page 8
$\odot$	Pointwise product of matrices, page 9
$\odot$	Pointwise product of tensors, page 11
$\otimes$	Kronecker product of matrices, page 9
$\times_k$	k-mode product of a tensor and a matrix, page 11

List of Figures

List of Tables

List of Algorithms